by the same author

FICTION
Loving Monsters
The View from Mount Dog
Gerontius
The Bell-Boy
Griefwork
Ghosts of Manila
The Music
Cooking with Fernet Branca
Amazing Disgrace
Rancid Pansies
Under the Radar

CHILDREN'S FICTION
Flight Underground
The House in the Waves
Hostage!

NON-FICTION
A Very Personal War: The Story of Cornelius Hawkridge
(*also published as* The Greedy War)
Mummies: Death and Life in Ancient Egypt
Playing with Water
Seven-Tenths
America's Boy
Three Miles Down
Marked for Death
Eroica: The First Great Romantic Symphony
Blackbird: The Untouchable Spy Plane

POETRY
Option Three
Dutch Alps

JAMES HAMILTON-PATERSON

EMPIRE OF THE CLOUDS

The Golden Era of
Britain's Aircraft

FABER & FABER

First published in 2010
by Faber and Faber Limited
Bloomsbury House
74–77 Great Russell Street
London WC1B 3DA
This new and revised paperback edition first published in 2018

Typeset by Faber and Faber Limited
Printed in England by CPI Group (UK) Ltd, Croydon CR0 4YY

A CIP record for this book
is available from the British Library

ISBN 978–0–571–34148–1

2 4 6 8 10 9 7 5 3 1

In memory of Squadron Leader W. A. Waterton GM, AFC*

and of the countless aircrew, designers and engineers
whose heroic work
in the dangerous early years of the Jet Age
made modern flying safe

They say Great Britain is still a first-class power, doing well and winning respect from the nations: and if so, it is, of course, extremely gratifying. But what of the future? That was what Lord Emsworth was asking himself. Could this happy state of things last? He thought not.

P. G. WODEHOUSE, *Blandings Castle* (1935)

Gentlemen, you must never, *ever* forget that all aircraft manufacturers are thieves and rogues.

SQUADRON LEADER EDDIE RIGG

Contents

Illustrations

The Fairey FD.2 going into a climb in March 1956
(Bettmann/CORBIS)

Short's S.A.4 Sperrin, August 1951 (Vic Flintham
Collection)

The Vickers Valiant B.1, the first of Britain's V-bombers
to go into squadron service (Science Museum/Science
& Society Picture Library)

An Avro Vulcan B.2 performing its celebrated leap into
the air on take-off, September 1959 (Popperfoto/Getty
Images)

Handley Page's HP.80 Victor prototype in July 1953
(Charles Edward Brown Collection/RAF Museum)

A formation of English Electric Lightning F.6s of No. 74
Squadron (Royal Air Force Official Photographer/
Imperial War Museum)

Two Blackburn Buccaneers in typical low-level flight,
1965 (Popperfoto/Getty Images)

The ill-fated English Electric/BAC TSR.2, prototype
XR219, autumn 1964 (BAE Systems)

A British European Airways' Vickers Viscount 630
prototype, July 1950 (Getty Images)

G-ANBC, a Bristol Britannia 102 in BOAC livery, July
1955 (Charles Edward Brown Collection/RAF
Museum)

The first of BOAC's Vickers VC10s, April 1962 (Getty
Images)

Bill Waterton, Gloster's chief test pilot caught in the late
1940s in characteristic pose (Courtesy of the Estate of
Bill Waterton)

Bill Waterton at 90, shortly before his death (Courtesy of
the Estate of Bill Waterton)

The Hawker Harrier's predecessor, the P.1127 FGA.1
Kestrel, demonstrates a vertical take-off, September
1965 (Getty Images)

Preface to New Edition

This new edition of *Empire of the Clouds* incorporates a good few corrections of detail for which I am indebted to sundry kind readers – most of whom were in the aero industry, RAF or Fleet Air Arm in the post-war period and who therefore had first-hand knowledge.

Since the book was first published in 2010 there have also been some notable changes to what is left of the nation's aviation scene. Two of Britain's most charismatic aircraft have flown for the last time: the Vickers VC10 and the Avro Vulcan XH558. Our entire fleet of 72 Harriers – already grounded by the disastrous Strategic Defence and Security Review of 2010 – was summarily sold off in 2011 to the United States Marine Corps for £116m, cheerily described by then-Defence Minister Peter Luff as 'a good deal both for UK taxpayers and the US government'.

Following the accident at Shoreham Airshow in the summer of 2015 when a Hawker Hunter T.7 crashed and killed eleven people who were not even spectators, a pall of uncertainty has hung over the future of air displays in the UK, despite their popularity, and it is still unclear how many of our few remaining airworthy Jet Age aircraft will be allowed up again, and for how much longer. Already the flying days of the one remaining Sea Vixen (XP924) seem likely to be over.

In the year the RAF celebrates its centenary, this steady erosion of the flying remnants of our post-war aviation heritage seems a good reason to publish a revised and updated account of what went wrong and what went right in Britain's meteoric aviation industry after the Second World War. In it will be found ample scope for pride in what we once pioneered, as well as remorse for what we allowed to happen.

Acknowledgements

My apprenticeship in Italy as a bee-keeper led me to believe that, in general, apiarists were the friendliest practitioners of any profession I had encountered. Since researching this book, however, I have had to conclude that retired test pilots present them with stiff competition. Without exception the ones whose brains I picked – often without notice and at inconvenient hours – were unstintingly kind and helpful. The undiminished enthusiasm they retain for flying and all things aeronautical resulted in an eagerness to answer my questions with patience and clarity. Here I am particularly indebted to Wing Commander J. A. 'Robby' Robinson, A. R. ('Al') Pollock, Mike Pearson and above all to John Farley, who has taken time and infinite pains that I should thoroughly understand a variety of technical matters, as well as generously putting me in touch with friends and acquaintances. Indeed, it was he who enabled me to track down the Watertons, from which contact the heart of this book has derived.

My researches in Canada into the late Bill Waterton's life and archive were pure pleasure, thanks to the great friendliness and hospitality offered by his widow Marjorie, his elder son Willy and family (Audrey and Bella), and his younger son John. I owe Willy an especial debt for his willingness to undertake the not always congenial task of digging out and going through his father's scattered papers and log books. Despite being a busy staff photographer on the local newspaper he could not have been more generous with his time, his thoughts about his father, and his keenness to show me as much as possible of the scenic Grey-Bruce landscape. The entire Waterton family bore with astonishing good grace the ordeal of having a total stranger descend on them and ask often intimate questions about Bill.

My stay in Ontario was made still more pleasurable by becoming acquainted with close friends of Bill Waterton's, the ex-test pilot Richard Bentham and his wife Lorraine, both of whom were wonderfully hospitable and informative. Since then, in the aviator's tradition of kindly enthusiasm already alluded to, Richard has been solicitous as well as eager that I should get my facts right, whether about his old friend or about aircraft and flying in general. I hope he and the Waterton family will feel their trust has not been misplaced.

Also in Canada, I should like to thank Willy Waterton's colleague Jim Algie for the invaluable recording of his interview with Bill Waterton in December 2003. The rarity of having this reclusive man's voice on record, and at relaxed length, makes his tape all the more precious.

In the UK I should like to thank Brian Riddle, the Royal Aeronautical Society's Librarian, for his assistance and advice; and Alistair Tucker for a most enlightening conversation about commercial aviation, his field of lifelong expertise. My old friend Michael Hall also proved a splendid source of reminiscences of flying the old Colonial routes, as well as of incisive political comment – the often heady distillation of a diplomatic career.

As usual I am greatly indebted to my editor Neil Belton, whose own interest in aviation was a decisive factor in this book's genesis, and his literary intelligence a still more vital influence on its final form.

My agents Andrew Hewson and Edward Wilson have provided me with tireless encouragement, counsel, and practical help of the most generous kind in the form of books and contacts and gossip. In particular my friendship with Andrew and his family, now stretching back decades, stands ever more clearly as fundamental to my sanity as to my scribblings.

I should also like to pay tribute to Trevor Horwood. This is not the first text of mine that he has copy-edited, and I trust it will not be the last. There is not a page of this book that hasn't benefited from his acute but always sympathetic revision.

Finally, I note that Acknowledgements sections seldom if ever mention books (rather than people) to which the author is indebted. Wing Commander J. A. 'Robby' Robinson's volumes of reminiscences about his life as an RAF pilot from the 1950s onwards provide a valuable and entertaining insight into how things were in that service during the worst of the Cold War. John Farley's *A View from the Hover* is a superb source of aviation common sense and experience. And anyone inclined to underrate the sheer imaginative prodigality of Britain's aviation and missile projects between 1945 and 1970 would do well to consult the scholarly and beautifully illustrated series of books by Tony Buttler, *British Secret Projects*. Many of the ideas were quite extraordinary and included designs that even today look futuristic.

PERMISSIONS

Grateful acknowledgement is made to Willy Waterton as the sole executor of his father's estate for permission to quote from Bill Waterton's *The Quick and the Dead* (Frederick Muller, 1956), and also to Charles Masefield for permission to quote from his father's autobiography, *Flight Path* (Airlife, 2002).

Similarly, I am indebted to the following: The History Press Co., for permission to quote from Robert Gardner, *From Bouncing Bombs to Concorde* (Sutton Publishing, 2006); Grub Street Publishing, for permission to quote from Neville Duke, *Test Pilot* (Grub Street, 2006) and Peter Twiss, *Faster Than the Sun* (Grub Street, 2005); Haynes Publishing, for permission to quote from Roland Beamont, *Fighter Test Pilot* (Patrick Stephens, 1987) and Brian Rivas and Annie Bullen, *John Derry* (Haynes, 2008); Martyn Chorlton of Old Forge Publishing, for every courtesy extended when sourcing his titles from abroad.

Every attempt has been made to trace the copyright holder of Roland Beamont, *The Years Flew Past* (Airlife, 2002), but with regrettable lack of success. I nevertheless acknowledge a debt of gratitude to this valuable source.

J. H.-P.

Introduction

At some point in the mid-seventies my friend and ex-tutor, the late Jonathan Wordsworth, offered me a job fund-raising for the Wordsworth Trust – or maybe it was the Dove Cottage Appeal. I was soon to be revealed as the world's worst fund-raiser, although consistent in that never did I raise more than I was being paid. However, until this fact became undeniable I was assigned a red Vauxhall Chevette and a secretary/assistant, a Wordsworth DPhil in her twenties. She and I were introduced in Oxford on the day we were to drive up to the Lake District for an official briefing in Ambleside. She struck me as earnest in a *Guardian*-reading sort of way; I evidently struck her as male, although I did my best to make my gender unobtrusive.

Somewhere up in the northern Midlands – I can see a moorland landscape with dark low clouds, probably in the Peak District – there occurred one of those epiphanic moments which take one by such surprise that all pretences are stripped away in a flash. From behind a hill a bare couple of hundred yards ahead a vision swept into view. 'Oh God! *Look!*' I cried involuntarily. It was an Avro Vulcan B.Mk 2 in camouflage livery, travelling not very fast and only a couple of hundred feet up, thanks to its terrain-following radar: almost close enough for its serial number to be legible. As if the pilot impetuously decided to put on a show for the tiny red car plodding along this empty road, it performed an almost vertical bank to port directly in front of us before going to full military power and climbing steeply away overhead at an astonishing angle. Against the dark sky the vast underside of the delta with its kinked leading edge was briefly revealed in all its ghostly majesty, the thunder of the four Olympus engines and their combined fifty-plus tons of thrust drowning out the Vauxhall's pitiful motor and setting up vibrations I could feel through the steering wheel as I watched the parallel trails of dark kerosene smoke heading up and vanishing into the cloud base. Shakily I slowed the car, repetitively mumbling something like 'Crikey!' as my gooseflesh subsided.

'I know,' my companion said with the first sympathy she had so far evinced. 'It's disgusting, isn't it? Just war, war, war. Endless boys' toys at the taxpayer's expense.'

I looked at her aghast. 'But ... surely ... it was so *beautiful!*' I blurted. That huge triangle against the sky like some monstrous angel trailing dark clouds of glory? It was *sensationally* beautiful: the most extravagantly charismatic aircraft this country has ever built. Doggedly I tried to explain that, yes, we're all against war, but you can't deny some of its machinery is downright sexy. That massive power, all those tons of metal just being blasted straight up into the sky as if they were so much thistledown? Sheer glory. You could feel it in your stomach as much as hear it. Surely you've got to be moved by the awesome prowess involved in designing and building something like that? The sheer human ingenuity? No?

No. My advocacy failed utterly and my new assistant spoke barely another word until we reached Dove Cottage, by which time it was clear we belonged to two radically incompatible species and I had unabashedly reverted to my native gender. The odd thing was that a second epiphanic moment occurred not long after this, and again with a Wordsworth connection. I was staying with Jonathan and his family in their ageless and remote little rented farmhouse above the Duddon Valley and we were walking one day up on Harter Fell. It was one of those comparatively rare idyllic Lakeland days when the sunlight falls unhindered from the sky with brilliant largesse as if to apologise for all the mist and rain and grey months. We were quite high up on the fell when the silence was broken by a strange bleating rumble. Suddenly there appeared from the shoulder behind us two SEPECAT Jaguars practising high-speed contour-hugging. They passed a bare fifty yards away but *below* us, so that we glimpsed the pilots' white bone domes behind their cockpit canopies as they flashed past in a drench of sound that rattled the hills before fading with that odd wailing quality caused by the complex echoes thrown back from the close, sheep-bitten slopes. The scent of burnt kerosene lingered faintly in the Lakeland air.

Once again I was reduced to schoolboyish excitement while Jonathan, who seldom voluntarily left the late 1790s, was genuinely pained and bewildered. His inner musing shattered, he attempted to overcome his distress with a quotation from his forebear: 'deep/ But short-lived uproar, like a torrent sent/ Out of the bowels of a bursting cloud . . .' But it didn't work as a jocular exorcism and he fell into gloom at the remorseless way in which the world is too much with us. Only I was left thrilled, although obscurely crestfallen for being so easily revealed as both juvenile and philistine. *But the world moves on*, I wanted to say (and probably did). The Lake District of the 1790s was no doubt full of wonderfully ethnic pedlars and leech-gatherers and ruined cottages, not to mention the music of humanity; but a couple of centuries later the hills are alive with the sound of Jaguars, whether in the air or in the car park down at Dove Cottage, and there's poetry in that, too, besides more money for the Archive. It simply requires the right poet . . .? No again.

I must try to be precise about this clash of sensibilities. It is made no simpler by so easily being able to put myself into a pre-industrial mood and bewail the vanishing of bucolic silence and unhurried communion with what we think of today as 'the natural world'. On the other hand I have always been moved by the science and aesthetics of flight and of particular aircraft, just as certain mariners have been by the sea and individual ships. Still, I can't deny that in my exultation at the two Jaguars storming by at our feet and blasting apart the immemorial hush there was also an adolescent component of rebellious glee. *To hell with the eighteenth century and all those visionary but often sententious pentameters! I am a creature of my own day and age, and that is* – the Jet Age! Childish, no doubt; but then aircraft have had significance for me from my earliest childhood.

At the time of the Coronation in 1953 I was eleven, and I and my friends were adjusting to our novel identity as New Elizabethans. Through children's periodicals such as *Eagle* and *Boy's Own* we were proclaimed the inheritors of a new order: a post-war Britain of amazing technological energy and inventiveness suitable for the new Jet Age. Being fervent plane-spotters, we walked about with permanent cricks in our necks from gazing heavenwards. The least sound or movement in the skies would instantly pre-empt our attention as we strained for a glimpse of one of the exotica Britain seemed constantly to be producing: experimental or prototype aircraft, most of which were destined never to go into mass production. The over-whelming impression we had was of a national industry at a peak of fertility and excitement. Scarcely a week passed, it seemed, without Britain claiming some new world record in speed or distance or altitude. Now that the near-mystical 'sound barrier' had been definitively broken there could surely be no further limit to aeronautical progress. For my friends and I the sky was our chosen playground. The rubber-band-powered propeller aircraft we used to launch into the wind now began to seem old-fashioned. To be properly up to date, the models we made from balsa wood kits now had to be powered by little Jetex jet engines that sent them hissing through the air, trailing a plume of pale, acrid smoke behind their fashionably swept-back wings. Our identification with what was literally 'leading-edge' technology was complete. We ran across a lot of fields, climbed a lot of trees and ventured into a good many back gardens in order to rescue our little plastic pilots glued beneath their clear bubble canopies as they manfully strove to emulate our flesh-and-blood heroes in the sky.

Today, more than half a century later and amid the industrial rubble of supposedly 'broken' Britain, it is not easy to convey plausibly the excitement of those times. The pages of *Boy's Own* were full of ads for a career in the RAF. This was the age of the jet, they said. Your future is in the air and *can start when you're fifteen* . . . Most of us were destined to get no nearer the military than National Service, and many were to escape even that. Yet at a fantasy level it remained a potent dream, like that of the day when we would be adults, and rich. In the interim we identified with the men who flew the new supersonic aircraft. None of the ordinary sporting heroes of the day – the mere footballers and cricketers messing about with balls and pieces of wood – had for us the test pilots' transcendental quality that went with *technology*: with the electrifying double boom that made parents curse and heralded the hair-erecting sight of a streaking aircraft pulling up into a vertical climb while filling the horizon with crackling thunder. We hero-worshipped the pilots, while the aircraft themselves were above all awesomely beautiful: stream-lined, violent, slitting the sky with surgical precision to reveal a dazzling – if quite undefined – future.

Such was a schoolboy's romantic viewpoint, necessarily naive and penetrated by dreaming. This befitted someone who, after being taken on his first flight (in a de Havilland Dragon Rapide for a 'circuit' of what was then still known as London Airport) had gone straight home and laid out some suitable pieces of timber on the garage floor in the hopes that they would somehow miraculously assemble themselves into a flyable version of that sturdy biplane. The actual world of funding shortfalls, complex military forecasting and politics that even then was dictating Britain's aviation future went quite unrecognised by us boys. The aircraft industry itself, though, was all too aware of the harsh realities that were steadily turning it into a dire mess. Yet those of us who went to air shows like Farnborough in the fifties contrived to keep our ignorance blessedly intact by means of a schoolboy patriotism. No matter how shaky the economic foundations may have been beneath the shows of glittering machinery, it was undeniably all *ours*. Every last rivet, every instrument, the brains behind the vision, the design, the manufacturing skills, the occasional gruesome mishap, the spate of world records – all were British to the core. In the next decade the English Electric Lightning scribbling its contrails across the sky within ninety seconds from brakes-off became an expression of British technology at its apogee. We could never have guessed how swift was to be the fall from that summit, nor to what degree the plunge would coincide with a diminishing of the nation's expectations, of confidence in its polity, of its technical abilities, even of its own self-esteem.

No one would seriously advance the thesis that Britain's morale in the last sixty-odd years has been pegged to the peak and fall of its aircraft industry. Yet in 1945 it was the UK's largest and most productive; and its decline has not been accidental but the result of economic forces and political decisions that have deeply affected all aspects of national life over the same time span. One could suggest that the story's riches-to-rags trajectory has faithfully kept pace with a good deal else in Britain that has scarcely been a cause for national pride. And maybe after more than sixty years' unbroken peace a good many people like my erstwhile Wordsworthian secretary will have turned eco-warrior these days, agreeing with her 'boys' toys' view of the Vulcan rather than with mine. (Yet the last flying Vulcan, a B.Mk 2 with the serial number XH558, which was privately bought and finally restored to airworthy condition in 2007, was a huge draw at air shows all over the UK where – to judge from newsreel footage of the spectators – small boys were every bit as enraptured by it as I was by its earlier versions.)

Is this, then, merely a nostalgic tale of a once-exalted national expertise that became unaffordable and has long since been overtaken by political and economic reality? At one level, of course it is. The world has moved on. We no longer build men-o'-war, either. The New Elizabethan era is now deader even than Wordsworth's eighteenth-century Lake District, which is painstakingly recreated for tourists. It came naturally to us schoolboys fifty years ago to mock the antediluvian British army general who in 1938 had predicted there would still be a vital role for the cavalry in any future war. Might those New Elizabethan engineers and pilots similarly cut laughable figures of dinosaurian pathos among today's children? No doubt. Yet for all the mismanagement of Britain's aircraft industry in the aftermath of the Second World War, it nevertheless did comprise prodigious talent, skill and inventive energy, not to mention individuals who daily risked their lives for it. If it is worth dipping into that history for reasons other than nostalgia, maybe we could start by wondering whatever had become of all those national high spirits, the dash and verve and daring.

For example, it is revealing to look at some of the air-to-air photographs of prototype aircraft being tested in the late forties and fifties. Inside the cockpits, their grinning faces turned towards the camera only yards away, are men often in their shirtsleeves, or wearing a jacket and tie, quite frequently bareheaded. Nothing better illustrates the gulf between the world they inhabited then and our own today than the way they dressed to fly. It is not just the informality that strikes us but the lack of *kit*. From the present day, ruled as we are by the endlessly repeated mantra of 'for your own safety and security', we look back at men carrying out a morning's worth of potentially lethal spinning trials wearing the jacket from their demob suits and with nothing on their heads but a dab of Brylcreem. In some of the pictures they seem not even to be wearing a parachute. The point is not that people today make too much of a fetish about kit; rather, that the change in attitude towards safety has had certain consequences. Nor is it to say that modern Britain has become too *tame*, exactly, especially for boys (although a growing interest in extreme sports is a good argument for the hypothesis), but that being increasingly apprehensive and risk-averse in matters of personal security percolates through everything else. The buoyant lift of adventure is too easily cancelled by the drag of Health & Safety Executive regulations, insurance, compensation claims, disability pensions and, well, social disapproval. Are today's Britons too collective and cowed for rugged individualism, let alone for civil disobedience? Too bored or lazy to acquire manual skills? These are imponderables.

In the meantime, though, we have a spectacular couple of decades to look back on when certain Britons took all sorts of wild risks on a daily basis, and a generation of small boys as well as millions of ordinary householders loved them for it. Along the way, I hope this book may also act as a modest first step towards restoring the reputation of one of Britain's (and Canada's) most courageous and influential test pilots.

I should add a caveat. This book is in no way intended to be a comprehensive survey of all the aircraft that Britain's post-war industry built and flew. I prefer to concentrate on those that made the greatest impression on me at the time. I also make no apology for paying less attention to commercial – as opposed to military – aircraft. Roughly seventy per cent of the industry's efforts were devoted to military machines, so there was always going to be an imbalance and I make no bones about tipping that imbalance still further. Britain's post-war output of light civil aircraft was almost non-existent; and as for airliners, I readily admit to being a good deal less interested in what I have always thought of as aerial buses.

Another personal bias – that towards fixed-wing aircraft – led to a shameful omission in the first edition of this book. There is in fact one British world air speed record that remains unbeaten to this day: that for heli-copters. In 1986 a Westland Lynx, G-LYNX, achieved the then-incredible speed of a shade under 250 mph. In May 2017 this remains the official FAI record for helicopters. Recently the Eurocopter X3 has managed 293 mph but it is a hybrid, with two propellers in addition to the rotor. That an absolute speed record like that of G-LYNX should have stood for over thirty years in such a stiffly competitive high-tech military field is quite remarkable.

More recently, in late 2010, as part of its Strategic Defence and Security Review, the new UK Government axed the Harrier and the Nimrod MRA.4. RAF chiefs had reluctantly decided the Tornado fighter offered broader utility than the Harrier GR.9 while they waited for the F-35 Joint Strike Fighter. BAE's Nimrod MRA.4, a prime example of grossly incompetent procurement and contracting, was £789 million over budget and nine years late. £3.5 billion had been wasted on endless fiddling with a flawed design while an entirely new aircraft could and should have been commissioned and built years ago.

1

Death at Farnborough

If you wished to choose a moment in the last sixty years of British history to reveal a country that was most unlike today's Britain, you could do worse than pick 6 September 1952. In technological prowess and social attitudes alike, the Britain of that Saturday over sixty years ago was effectively a different country altogether. As for the best vantage point from which to view things on that day in the distant past, it might be instructive to choose the Society of British Aircraft Constructors air show at Farnborough, Hants. After the two immediate post-war years when the show was held at Radlett, the SBAC had decided on Farnborough as the most appropriate place in which to hold their industry's annual showcase week. The Royal Aircraft Establishment's home base, with its international reputation as a cradle of aeronautical research – not to mention its long runways and experienced air-traffic controllers – seemed the right setting in which to display the all-British aircraft they hoped to persuade foreign air forces and airlines to buy. The first four 'trade' days were given over to wooing such potential customers; the last two were open to the general public.*

---

* Two significant changes since then are that in 1962 the SBAC air show became biennial and international, alternating with the Paris air show and being held in mid-July in even-numbered years. And as from 1964, SBAC stood

On this particular Saturday the public were out in force: some 120,000 of them, wandering among the aircraft parked on static display and anticipating the afternoon's flying demonstrations. Periodically they stopped to consult the gaudy Official Programme (price one shilling), whose cover showed a futuristic blue swept-wing jet streaking across a bright yellow background. Occasional announcements over the tannoy system blatted off the 'Black Sheds' and lent a fairground touch to the festive scene, as did the white canvas walls of the scattered booths occasionally paunching in the light breeze. There were only a few tents belonging to individual exhibitors as well as the SBAC president's large marquee overlooking the main runway. The weather was greyish, as it had been all week, but the clouds were high enough to allow good views of the aircraft that would be put through their paces in the programme after lunch. In any case it would have taken far more than indifferent weather to dampen the air of anticipation. The crowds, who even on a public day included VIPs and military representatives from ninety-four countries, had come to wonder at the sheer variety of new aircraft the British aviation industry was producing with such apparent fertility and panache. Farnborough '52 was to be the year of the delta: a novel wing shape widely believed to be best for supersonic flight and a progression from the swept-back wings hitherto thought indispensable for breaking the sound barrier (which to most people's eyes still looked exotic enough a mere seven years after the

---

for Society of British Aerospace Companies, as if in prescient recognition that the days when Britain actually *constructed* aircraft were drawing to a close.

war, when virtually all wings had been straight). On this public Saturday three different delta types were to be flown: the little experimental Avro 707s, the immense Avro 698 that had grown out of them and which would shortly be known as the Vulcan bomber, and Gloster's prototype Javelin fighter.

Exciting though the prospect was of these new shapes in the sky, what had really drawn many in the crowd was the chance to see and hear the sound barrier being broken. It is hard in the post-Concorde twenty-first century to convey how gripped public imagination was in the early fifties by the idea of supersonic speed. It had come to symbolise the last word in advanced technology and personal bravery, and for a period was to become the height of fashion at air shows, much as 'looping the loop' had been forty years earlier. In 1952 everyone at Farnborough would have known that the first person officially to break the sound barrier had been the US test pilot Chuck Yeager, in an American rocket plane in 1947. But they also knew that the first Briton to break it in a pure jet (and British) aircraft had been John Derry, who had done so the following year – four years ago to this day – in a de Havilland 108 Swallow. This afternoon he would be doing it again, but in the same company's more advanced DH.110. What was more, Hawker's equally famous chief test pilot Neville Duke would also be flying faster than sound in a Hawker Hunter. Both pilots had gone supersonic on the previous four afternoons for the trade spectators; but although in those days the sky over southern England quite often echoed to sonic booms (provoking furious letters to the newspapers by the sort of correspondents who write furious letters), few people ever managed to see the aircraft responsible since it was usually out of sight before the noise made them look up. Today they would have a chance of watching the entire thrilling process.

Part of the attraction lay in knowing that the speed of sound still represented a limiting factor in the aerodynamic knowledge of the day. The sense was of men and machines flying at the very edge of what was understood and could be built. An aura of heroism and danger hovered about the entire enterprise, helped along by the judicious slight exaggeration in David Lean's newly released film *The Sound Barrier*, even then playing to packed cinemas across the land. The film contained obvious echoes of Geoffrey de Havilland Jr's death over the Thames estuary in September 1946 when the DH.108 he was testing broke up in the air, not to mention the lingering suggestion that he might after all have been the first man anywhere to exceed the speed of sound seconds before it cost him his life. It was undeniable that many people at Farnborough this Saturday had been unconsciously drawn by the chance of witnessing something going wrong, just as they were to motor races. The prospect of real-life drama and spectacle was particularly enticing at a time when there were still fewer than two million black-and-white television sets in the country. Aviation in 1952 was not quite as disaster-prone as in earlier days, but even so there had been eighty-two British flight test fatalities since January 1946.[1]

The men who assembled in their flying suits in the pilots' tent and enclosure over by the control tower embodied an appropriate mystique. Most were test pilots of one sort or another, whether flying for the Ministry of Supply at research establishments such as Boscombe Down some forty miles away on the edge of Salisbury Plain (also home of the world-famous Empire Test Pilots' School), or for the RAF, or for private aircraft manufacturers like de Havilland, Gloster or Hawker. They were an elite and many were already household names. John Derry, Neville Duke, Roland Falk, John Cunningham, Bill Waterton, Roland Beamont, Peter Twiss, Mike Lithgow . . . the names were as familiar then as those of many footballers are today, but they carried an extra voltage because they had nearly all flown in the recent Second World War and were authentic heroes. Those who hadn't been in actual combat had been test pilots flying experimental and prototype aircraft, a no less dangerous way of life. This was only one of several important ways in which they differed from today's celebrities (a word that itself belonged to a future of mass TV audiences). The test pilots of 1952 risked their lives for about £40 a week if they flew for a company like de Havilland, and half that if they flew for the RAF or the Royal Navy.

In those days only British aircraft were shown at Farnborough, a policy that bolstered national pride while allowing taxpayers an exclusive view of where their money was going. Not all the aircraft on show were ready to fly, and occasionally less secret prototypes and even mockups could be found on static display still smelling of hastily applied paint. Each year at the SBAC show there were dark rumours that by no means all the new aircraft that were flown were strictly ready for it. It was hinted that certain companies would urge their test pilots to demonstrate them as impressively as possible even when the exact limits of the aircraft's capabilities were still uncertain. That there might be something in these rumours can be intuited between the lines of an editorial in *Flight* magazine that appeared a short while later, which gave an emphatic view of how matters stood: 'To sell aircraft abroad is a most (if not *the* most) important single objective of the aircraft industry – both in its own interest and in that of the nation,' it began, evidently overlooking that a still more important objective might be to build first-class aircraft before worrying about foreign sales. A nicety too far for *Flight*, perhaps, given the grandeur of its vision a bare nine months before the Queen's coronation. It continued, 'The parallel of the growth of shipping in the first Elizabethan era and the signs of another great transport industry building up in the reign of Queen Elizabeth II makes an irresistible appeal.'[2] In short, behind the flying display's meticulously co-ordinated timetable at Farnborough (fast jets were allotted five minutes and slower commercial aircraft up to eight), and behind the upbeat tones of the loudspeaker commentary, pilots might be subject to a good deal of commercial and political pressure to 'put on a bit of a show' for Britain.

In addition to the three delta-winged aircraft advertised to fly, there would be a chance to see thirty-five different all-British aircraft. These included the vast ten-engined Saunders-Roe Princess flying boat as well as the Bristol Britannia, the big new propeller airliner that it was hoped would sell well to the world's airlines and help fulfil *Flight*'s vision of an air transport industry befitting the new Elizabethan age. In the meantime there were plenty of aircraft parked on the hardstanding for spectators to walk around and examine at leisure. These included two de Havilland Comet airliners, a Bristol Britannia in fresh BOAC livery, a hulking Blackburn Universal transport, the Short S.A.4 used for testing jet engines at high altitudes, four Canberra jet bombers, some Meteor, Vampire and Venom jet fighters and much else besides. Many of the larger aircraft were open so that enthusiasts could climb inside, sit in the cockpit and fantasise while revelling in what one pilot memorably described as 'that haunting aircraft smell . . . composed of a mixture of odours, of doped linen, rubber, high octane petrol and a faint overlay of vomit'.[3] He was actually describing a Second World War-vintage Avro Anson, but the smell of these more modern aircraft on display wouldn't have been much different, merely adding a waft of fresh hydraulic fluid in place of the dope and, in the jet- and turboprop-engined planes, the scent of paraffin (as kerosene was generally known in Britain): that hallmark fragrance of the Jet Age.

Although gas turbines would run on almost anything that could be mixed with air and burnt (from hydrogen gas to coal dust), aircraft jet engines were designed to use kerosene. Not only was this readily available, it was cheaper than aviation-grade petrol and offered the big advantage of being far less flammable: a characteristic that the navy particularly liked after its wartime experiences of the appalling fires that had often resulted when its aircraft carriers were attacked. By 1952 Farnborough Week's characteristic smell was the warm drift of burnt kerosene, reminding many spectators of the friendly odour of the Valor oil stoves with which they still heated their homes.

Not long before lunch a de Havilland Dove light transport could be seen taxiing out for take-off. After its departure the day's commentator, Charles Gardner, announced over the tannoy that it contained John Derry and Tony Richards, respectively the pilot and observer of the DH.110, who were having to fly back to de Havilland's base in Hatfield to fetch a replacement aircraft for their afternoon's display. It appeared that the starboard engine of WG240, the sinister but beautiful all-black aircraft in which Derry and Richards had been breaking the sound barrier all week, was defective. Charles Gardner told the crowds he very much hoped Derry and Richards would be back in time with the spare 110 to perform as planned, but apparently there was only one compressor starter available for both aircraft, and this was on a lorry being driven from Farnborough back to Hatfield at that very moment. Fingers crossed, in short, in recognition that the de Havilland 110 was definitely one of the most glamorous aircraft at the show and one of the main draws. It was designed as a two-seat interceptor for the RAF: a large aircraft with swept wings and a bubble canopy over its cockpit oddly offset to port. In place of a conventional tail, two narrow tubular booms extended behind the short fuselage, each with a swept-back fin at its end. The tops of these fins were joined by a rectangular tailplane poised safely above the efflux of the two jet engines. Anyone familiar with de Havilland's much smaller twin-boom jet fighters, the Vampire and the Venom, would instantly have recognised the 110 as being from the same design stable. In its all-black night-fighter paint scheme, WG240 had been thrilling the crowds with its sonic booms on the previous display days, but now engine trouble had rendered it unflyable. Hoping for the best, the spectators turned their attention back to lunch: typically a picnic of sandwiches with a Thermos flask of tea often eaten aboard the coach in which they had arrived.

As soon as the flying began, though, all thought of disappointment was forgotten. From the direction of the Solent the great Princess flying boat appeared. It had flown for the first time only a fortnight earlier, preceded by a barrage of patriotic superlatives in the press and newsreels about how it was going to ply the South Atlantic routes for the British South American Airways Corporation (even though BSAA had yet to order one). Passenger flying boats had become familiar before the war, especially those serving Imperial Airways' African and Far Eastern mail routes. Their sheer size had meant the provision of a lounge bar on the upper deck, but there the luxury ended. They were very slow and unpressurised so they couldn't fly above the weather, and apart from being bumpy it was often cold enough in the cabin to freeze the water in a passenger's glass. They also made frequent refuelling stops, sometimes having to land on choppy water. In difficult conditions it might take three-quarters of an hour for the flying boat to be safely warped to its moorings, and in the meantime those passengers who hadn't already succumbed to air sickness were often sea sick. Embarking and disembarking was via launches, on to whose bobbing decks passengers occasionally had to jump. Obviously, flying boats on these Empire routes did offer a huge saving in time over steamship travel and the air of modernity this gave them added greatly to their cachet. Still, their image was generally a good deal more stately than the experience they actually afforded. What was more, during the war a great many new airstrips had been built in far-off places, which tended to neutralise one of the flying boat's major pre-war advantages.

Saunders-Roe's huge new Princess was nevertheless designed to usher in a modern era of flying-boat travel, even though its passengers would presumably still need to be ferried to and from the aircraft in launches. This time their comfort on two decks was promised to be 'sybaritic'. However, this might have been difficult to imagine from the way in which Saunders-Roe's chief test pilot, Geoffrey Tyson, had displayed his employer's majestic new airliner earlier in the week. Tyson was a pilot whose flying style, while brilliantly accomplished, could not have been described as sedate. He had once flown a Tiger Moth biplane across the English Channel upside down all the way, and in 1948 at Farnborough had made an instant hit with the crowd while displaying Saunders-Roe's SR.A/1. This was the world's first jet flying-boat fighter, designed late in the war for service in the Pacific arena when the Japanese had been threatening landing strips and a shortage of aircraft carriers was feared. In the event there were enough carriers and the SR.A/1 never went into production, making its last appearance moored briefly to London's South Bank during the Festival of Britain in 1951. In 1948 Tyson had flown it at Farnborough as though it were a Spitfire, hurling it about the sky and even flying it inverted at low level: possibly the first time in history that anybody had ever deliberately flown a flying boat upside down.

And now, earlier this week, he had done something similar with the gigantic Princess, bringing it past the stands at Farnborough in an almost vertical bank, one wing tip perilously close to the grass. Years later it was revealed that he was travelling so fast that a glitch in the powered control system had almost prevented him from rolling back upright. It had cost him sustained physical effort to haul the massive machine back onto an even keel, having been for a long moment only feet away from an apocalyptic disaster.[4] What was never explained was why Tyson had ever thought low-level stunting was the best way of displaying his company's new airliner, especially one expressly designed to attract customers with its stateliness and passenger comfort. But that was display flying for you: there had to be a wow factor.

So here was Tyson on the Saturday, maybe chastened into making an ordinary low pass over the heads of the crowd. Despite the commentator's shouted accolade in which he listed the Princess's record-breaking vital statistics and future prospects as it droned impressively over, one wonders how many present realised that this graceful flying hulk was already obsolete. If so, the same cynics might have had a similar unpatriotic thought two years earlier when the Bristol Brabazon had made its Farnborough debut. After that gigantic airliner's maiden flight the Pathé News commentator had confidently predicted, 'The Brabazon will lead the world as Queen of the Skies,' although truth to tell not a single example had yet sold – and nor would one. Neither would the Princess and its two sister aircraft, equal victims of a complete absence of market research. It turned out that nobody needed flying boats to go to South America, and not even the military wanted them as troop transports, so all three were eventually mothballed and in due course broken up. It is open to question how many of the Farnborough crowd in 1952 thrilling to the sound of the ten Bristol Proteus engines low overhead would have entertained any such gloomy prognostications. In fact there was a good deal of spontaneous clapping and cheering as Britain's latest great white elephant lumbered away to its moorings at Cowes.

From then on that afternoon there was scarcely a moment when nothing of significance or interest could be seen in the cloudy skies above Hampshire. Charles Gardner's excited voice drew the spectators' attention to the approach of the deltas, and suddenly there they were: the huge white Avro 698 flanked by the tiny darts of the blue 707A (flown by the one-legged Jimmy Nelson) and the bright red 707B. The two little aircraft were one-third-scale versions of the larger model and had been designed to test the characteristics of the delta shape. The first Avro 707 had appeared at Farnborough three years earlier in 1949 but shortly afterwards had crashed, killing the company's deputy chief test pilot Ric Esler. The reasons for that accident were never fully clear. There were no sophisticated 'black box' flight data recorders and ejector seats were few and far between. Since then the 707A had been built, together with its variant sister the 707B, and useful data had been obtained from them about the characteristics of delta wings. The flying proof of this was even then approaching over the airfield's northern boundary in the guise of a vast triangular sail trailing dark smoke from its four engines. This was an early public glimpse of the 698, the prototype Vulcan bomber. It had left the ground for the first time a mere week earlier in the hands of the Avro test pilot, Roly Falk. He was another extraordinary individual who had secured his post partly by being an exceptional and experienced pilot but also by having the instinct and talents of a salesman. Not for him the rumpled hair and oil-stained flying suits of some of his colleagues. Falk's trademark was to fly in a grey pin-striped suit and tie, which with his neatly groomed thinning hair made him look every inch a businessman.[5] In future years this would enable him to perform a dazzling display in front of potential customers and then step straight out of the cockpit looking immaculate and go in to lunch with fellow pilots like Prince Bernhard of the Netherlands or King Hussein of Jordan with not a hair out of place.

Today was only the fourth time Falk had flown the 698, which bore the serial number VX770, but he was not going to let that stop him putting on a show. A year or two later at Farnborough I was lucky enough to witness a legendary moment in British aviation when he actually *rolled* the Vulcan bomber off the top of a post-take-off climb with complete panache: a mere prelude to a display in which he threw it about the sky like a fighter. The sheer noise and spectacle were so physically and emotionally overwhelming I must have cringed involuntarily at the prodigious low-frequency vibrations. 'Scared?' my father asked with slight mockery. Not *scared*, exactly; just rendered helpless by a majesty even the Second Coming couldn't hope to emulate.* But in 1952 the prototype wasn't quite ready for that sort of treatment.

Had the Farnborough crowd but known, the new aircraft was far from finished. Because the Bristol Olympus engines designed for the Vulcan were not yet ready, VX770 was fitted with four less powerful Rolls-Royce Avons. Apart from that, quite a few of the cockpit instruments had yet to be installed, the second pilot's seat was missing entirely, and some large fairings on the undercarriage (which had fallen off the previous week during the maiden flight and hadn't yet been replaced) required redesigning. Undaunted, Falk came in very low and quite slow, then pulled a steep climb and an almost vertical bank 200 ft above the dignitaries and spectators. The sight of this monstrous white ghost tilted vertiginously overhead practically at arm's length brought involuntary gooseflesh to many. Despite the aircraft's overwhelming star quality such a display was not without its critics. *Time* later quoted 'a top US plane manufacturer' as saying, 'That pilot ought to be shot. He risked the lives of dozens of the top aviation brains in the free world.'[6] Half a century later a retired Vulcan test pilot hinted that Falk had known exactly what he was doing. In 1952 there was a good deal of scepticism in official and political circles about these new delta wing shapes and Avro, already facing fierce competition from Handley Page's Victor

---

* Falk did this twice before he was told to stop – not on safety grounds, since it was a low-*g* manoeuvre (i.e. it did not generate gravitational forces high enough to endanger the airframe) and the aircraft had power to spare, but because the SBAC itself unsmilingly considered it 'inappropriate behaviour for a bomber' (*The Aeroplane*, January 2009, p. 10).

bomber with its more conventional outline, may have urged Falk to put on an impressive demonstration of the aircraft's performance on the grounds that if it was a public sensation it would be that much harder for Westminster to cancel the Vulcan programme.[7]

After VX770 had picked up its smaller escorts once more and headed away a de Havilland Comet made a steep rocket-assisted take-off, presumably to assure potential customers for the world's first jet airliner that if need be it could shorten its take-off run to suit small airports. Vickers's chief test pilot 'Jock' Bryce flew a Vickers Viscount past in near silence with three of its four Rolls-Royce Dart turboprops feathered (i.e. the propeller blades of its stopped engines were turned edgeways on to the airstream so as to offer the least resistance). It then made a climbing turn still under the power of a single engine, an impressive demonstration of the inbuilt safety margin of this potentially great small airliner. Then without warning a silver Supermarine Swift blasted by on a high-speed run that had spectators covering their ears. This was the swept-wing jet fighter whose prototype was even then appearing in *The Sound Barrier* film under the name Prometheus. Unfortunately, its celluloid counterpart was promising a much more dazzling future than the Swift was destined to enjoy. Designed and built by the same company that had produced the Spitfire, it never really lived up to its noble heritage despite briefly holding the world's air-speed record, and was something of a disaster for much of its short RAF service. But at Farnborough in 1952 one might not have guessed this as it made screeching passes at 700 mph before boring up in a steep climb into the clouds, leaving the spectators pleasurably stunned by the noise. The next display supplied a complete contrast and accorded with British taste by ensuring that an awe-inspiring moment was immediately undercut by the ridiculous. In this instance Ranald Porteous gave one of his skilful exhibitions of crazy flying in a little propeller-driven Auster Aiglet, ending up with his crowd-pleasing one-wheel landing.

The less diverted might have noticed that in the interim the de Havilland company's Dove had returned, having left John Derry and Tony Richards at Hatfield biting their nails while waiting for the compressor to turn up on the lorry to start up the spare DH.110. Finally it arrived at Hatfield, pulled in beside WG240's sister aircraft, WG236, and soon had both engines going. WG236 was painted a silver-white that made it look somewhat less menacing than its all-black sister, maybe even faster and more futuristic. Glancing at his watch, Derry calculated that they could just make their appointed slot in the air show's crowded schedule. He quickly ran the usual cockpit checks, taxied out and took off for Farnborough, which was only about five minutes' flying time away in this high-speed new jet. He had no real need of an observer for the brief daylight run but young Richards was extremely keen and wouldn't have been left out for anything. They had been a team for some time now while testing the 110 and had come to rely on each other. In any case Derry was someone most people would have given their eye teeth to fly with. He was frankly glamorous in the casual, understated fashion of an ex-RAF pilot with a distinguished war record who still looked like a captain of school cricket.

He was appropriately modest and charming and dedicated to his job, as well as representing the new breed of younger test pilots who were alive to the more advanced technicalities involved in flying supersonic jets and who tended to be less flamboyant than professional and methodical.

He now climbed to 40,000 ft, from which altitude he would dive in order to begin his display by breaking the sound barrier. It was easy enough to do this; the difficult part lay in arranging his flight path so that the bangs were audible in the right place. Earlier that week he had dived the black 110, done his fast run in front of the crowd and ended with some impressive aerobatics designed to display the aircraft's manoeuvrability. It had been an impeccable show except that the bangs were inaudible at Farnborough, although they had jolted folk in Hartley Wintney a few miles to the north-west. The problem in aiming sonic booms was that they might have a comparatively narrow 'footprint' only a mile or two wide depending on the aircraft's height, its angle of dive and even the weather.

Down on the ground, Charles Gardner's voice booming over the loudspeakers drew the crowd's attention to a patch of blue sky across which a silver speck could be seen speeding in rapid descent. On Observation Hill, the rising ground well back from the main runway that offered the best view (and which today is covered in hospitality suites), the tightly packed spectators craned their necks to follow it. They saw two transient white puffs of condensation appear behind the 110: the typical shock waves of an aircraft going supersonic. Not long afterwards a massive triple explosion shook the airfield and jolted people into clapping and cheering. This was what they had come for, after all: the violent music of a new technological era, the future made audible. All eyes were on the aircraft as Derry pulled it out of the dive and sent it scorching across the airfield.

As the crowd watched it out of sight the loudspeaker commentary said something about the aircraft and the stunts the pilot would make on his return. In the cockpit Derry would have been well aware of the time element. He and Richards now had a mere four minutes left in which to put the 110 through the rest of its paces before Neville Duke went up to make some sonic booms of his own in the Hawker Hunter. Derry banked round behind the airfield before turning back towards the crowd, travelling slower than after his supersonic run but still at about 650 mph. The crowd watched Derry head towards them and then pull up a little, presumably about to perform an upward roll. In the same instant the 110's outline suddenly blurred and they saw two small black objects detach themselves and soar upwards in a great arc, heading towards them. It took a second or two for people to register that the entire aircraft had disintegrated in mid-air. Some lighter alloy panels were already fluttering to earth while the larger pieces of cockpit, inboard wing sections and the booms went tumbling through the sky. The stunned silence was broken by Gardner's loud cry over the tannoy: 'Oh my God, this was never meant to happen!'

Most people's attention was probably still fixed on the recognisable parts of the aircraft as it broke up, especially the largest, consisting of the twin booms and inner parts of the wings. This pitched up, seemed to stall, and then at an almost leisurely pace fell on a mobile generator near the airfield boundary. But the real danger was from the two black objects describing their arcs towards the crowd: the Avon engines, searingly hot and each weighing a ton. In the two seconds' silence following Gardner's shocked cry the twirling engines could be heard whistling faintly as they left faint trails of smoke. He could see from his vantage point what was going to happen and helplessly shouted again, 'Look out!' But no one could run anywhere. One engine thudded safely into waste ground beyond the perimeter fence while the other plunged four-square into the packed spectators on Observation Hill. In the same moment the truncated cockpit section with Derry and Richards still strapped in their seats smashed into the ground beside the main runway not far from the president's marquee, bursting into fragments that injured several of the front-line spectators. A faint scent of kerosene vapour hung over the scene.

What all witnesses seem to remember most vividly is the extraordinary silence that followed. It lasted even as people went to help the dying and injured. Some said it was like a reversion to the recent war when, after an air raid, survivors were often too dazed to cry out, or else too busy with rescue. Derry's wife Eve had watched the whole thing from the pilots' enclosure a few hundred yards away. She was standing between two of her husband's de Havilland colleagues, John Wilson (who had flown the Dove) and Derek Taylor (who was due to demonstrate a Heron later that afternoon). She, like her husband, was not easily panicked. She just said quietly, 'There's no hope, is there?' and Derek replied, 'No. None.' They had all been in the trade too long for illusions.[8] So had Bill Waterton, Gloster's Canadian-born chief test pilot, who was also standing in the little group. He was due shortly to demonstrate the prototype Javelin, the heavy twin-jet two-seat delta fighter that was viewed as the DH.110's rival. Derry was a friend; and only a few weeks previously they had spent the day together when Derry had been contemplating a possible job with Gloster.

First-aid teams had arrived swiftly on Observation Hill to deal with the dead and injured, who included several children. At the same time an ambulance took away the bodies of John Derry and Tony Richards. When I was back at school barely a fortnight later the egregious Wetherby, a plump lad who gave himself airs because his father drove a flashy black Austin Atlantic convertible, gained brief notoriety by claiming that not only had he seen the whole thing but that John Derry's body had landed only feet away from him. 'What was it like?' we naturally wanted to know. 'All horrible. Sort of like summer pudding,' was all he would say, but we none of us knew if he was telling the truth. It turned out he really had been at Farnborough on the Saturday concerned; but still, knowing Wetherby as being fonder of his inner man than he was of aircraft-spotting, we suspected he had been hundreds of yards away stuffing himself at the ice-cream stall. However, we did think he probably glimpsed some fairly unpleasant sights on that notorious afternoon. In those days there was no such thing as 'counselling', not that he appeared in need of any. Within a week or two his star had waned until even Wetherby appeared bored by his own story.

Meanwhile, forty teenaged technical apprentices from

RAF Halton who had come by coach for the day were summoned to form a cordon around the carnage on Observation Hill. It was soon apparent that twenty-seven people had been killed outright and over sixty injured, many gravely. It was a scene that has remained indelibly in the memories of witnesses who described it fifty years later, many of whom were children at the time. They remembered bodies being thrown into the air that were later covered with sheets of blood-soaked newspaper. They remembered seeing one man whose trouser legs had been shredded by people trampling backwards in panic as the engine came hurtling towards them. Someone recalled flakes of silver paint twirling down from the sky like unseasonal snow and settling on the yellow and blue programmes scattered in the mud. And one of the young Halton apprentices remembered having to maintain the cordon for many hours until it grew dark, and then finding that his coach party no longer had a driver because he was one of the dead.[9]

But people also remembered that after only as long as it took to clear the worst of the debris off the runway, the show went on as advertised. Bill Waterton sadly left the abruptly widowed Eve Derry in order to climb into the Gloster Javelin, an aircraft he knew in his heart was no safer than the rival DH.110, and show it off even as ambulances were still converging on the scene of the disaster. Derry's friend Neville Duke had already taxied out and was now lined up for take-off in the duck-egg green P.1067 Hawker Hunter prototype in order to perform his allotted part of the programme. People remember the crowd as quiet and apprehensive. To nearly everybody it seemed no coincidence that the break-up of Derry and Richards's 110 happened just after they had broken the sound barrier, and here was Duke about to attempt the same thing. Suppose the shock of the accident had taken the edge off his reflexes? Duke later remembered exactly how he had felt:

There was nothing I could do. I stood discussing the accident . . . until it was time for me to get into the Hunter. I felt very sad at losing another good friend – so very many had gone during the war . . . and now John Derry.

Soon I had to stop thinking about them. It was time for me to go off, but there was a bit of delay while the wreckage of the 110 was being cleared from the runway.

'Please keep to the right-hand side of the runway on take-off and mind the wreckage,' I was told by control tower over the r/t.

'Roger.'

'Are you going to climb and do a bang?'

'That is roger.'

'Will you soft-pedal your display over the crowd, please?'

It was a lovely day for flying. At 43,000 feet over Odiham I could see the airfield clearly. The cockpit was quiet and warm, everything was in first-class order. It would be untrue to say that I was not disturbed and worried by John's death. I reflected that so little is known of supersonic flight; perhaps it could have had something to do with the accident. Then it was time to dive. The Hunter did its stuff perfectly, the bangs were heard by the crowd at the display. When I landed I could see the ambulances still in the area where the DH had broken up. I hoped that many people had not been hurt.[10]

Without knowing the type of man Duke was, the period in which he was brought up and his fighter-ace war record, it would be easy to mistake laconic for heartless. He certainly had his critics at the time who said he should never have done his display, that the rest of Saturday's air show ought to have been cancelled. But the majority approved and were much moved by the gesture, seeing it as his homage to an old friend and colleague, a gallant salute to a fallen comrade. It was perhaps more comprehensible to those who could remember active service and how in war there was seldom time to grieve. Just getting on with things was its own form of quiet tribute. Duke's courage was recognised for what it was by many, including Sir Winston Churchill, who sent him a note the next day. It read: 'My dear Duke, It was characteristic of you, and of 615 Squadron, to go up yesterday after the shocking accident. Accept my salute.'[11] An avalanche of letters and telegrams arrived for Eve Derry and their two young daughters, as well as at de Havilland's. There were telegrams from the Queen and the Duke of Edinburgh and headlines for days in the world's press. Britain's own newspapers sounded a note of national mourning while keeping up titillating speculation for those of superstitious bent. Had the first Briton to break the sound barrier in a British aircraft done it once too often? How could the date not be significant, seeing that it was four years *to the very day* since Derry's first supersonic flight?

By the time the editorials began to appear the preliminary toll was known: twenty-nine dead including Derry and Richards, sixty-three seriously injured. It was the world's worst air-show disaster to date. Yet the following day, Sunday, and despite pouring rain, 140,000 spectators turned up at Farnborough to see the closing of the show. Doubtless not a few of them were ghouls. As for the editorials, they tended to be written in much the same stiff-upper-lip style as Neville Duke exhibited. The next issue of *The Aeroplane* (12 September 1952), ever-mindful of the industry's primary commitment to commerce, reported: 'After a few minutes' delay the show went on and to many the sonic boom announcing Neville Duke's arrival in the Hawker Hunter was the best possible affirmation of the country's determination to press on with the work for which so many have given their lives.' On the same date *Flight* wrote of Derry and Richards: 'Such men are the pathfinders of the supersonic age upon which, for better or worse, we now enter. The inquest jury's verdict – that Derry and Richards "died accidentally in the normal course of their duty" – must be their epitaph.' The word 'duty' implies that *Flight*'s editor had forgotten the two men were not actually in the military but working as civilian employees of a private company and thus had been doing their salaried jobs. But the Korean War was still being fought and the Cold War becoming frostier; and where high-performance jet fighters were concerned it was anyway easy to conflate the civilian and the military, and both with the imperatives of the recent war. In those days 'duty' still had deep moral echoes.

The accident cast a dark pall over the de Havilland company even as their investigators tried to discover exactly what had happened. It felt uncomfortably close to a repeat of the accident in 1948 that had killed Geoffrey de Havilland Jr when his DH.108 broke up. His father was scarcely less affected by Derry's and Richards's deaths.

Many cinema-goers seeing *The Sound Barrier* presumed that the character of Sir John Ridgefield, the ruthless aircraft company tycoon determined at all costs that his new jet plane will go supersonic, was to some extent modelled on Sir Geoffrey de Havilland. It turned out that neither man was heartless so much as strong and silent; but adding to Sir Geoffrey's grief for his chief test pilot and observer was the humiliation of having another of his aircraft disintegrate in mid-air, this time in the most public of all trade arenas and at awful cost to so many spectators. It also came at a delicate moment when the company was hoping to make inroads into the US market with their pioneering Comet airliner.

The company appealed for any film that might have been shot of the accident, and someone duly came forward with footage he had taken that contained the entire incident. In slow motion the film clearly shows the sequence of events as the 110 comes towards the camera in a left-hand bank. In little more than a second the outer panel of the starboard wing disintegrates, causing the port wing to rear up and lose its own outer panel under the strain. The aircraft pitches violently upwards and the cockpit section breaks off, followed by the tailplane as it is hit by debris. As Derry was pulling 4.5g in the turn, the wing failed at a point where a shear buckle had already begun to form, rendering it less than half as strong as it ought to have been. The proximate cause was flutter: vibration that could be violent enough to render a pilot temporarily blind and so radically fatigue airframes that main spars could fail and entire wings and tails break off within seconds. The 110's wing had simply been flexing too much: it was a design fault. Derry had done nothing he hadn't done countless times before in the same aircraft, but he had done it once too often. It had nothing to do with the sound barrier and everything to do a fatigue problem that had gone unnoticed. There were no computer simulations in 1952. If one was looking for irony, it could be found in an article that Derry himself had written on test flying for *The Times Survey of British Aviation* published only days earlier to coincide with Farnborough week. Incorrectly captioned as 'The first British pilot to exceed the speed of sound,' Derry wrote: 'The actual risk is not, as is commonly believed, primarily one of structural failure but of losing control over the aircraft or its moving surfaces.' True enough; but it was a brave thing to write at a time when structural failure was by no means rare and aircrew fatalities following it a commonplace.

One immediate consequence of the Derry disaster was a change in the regulations governing air displays. From then on, no aircraft was allowed to turn towards the crowd unless it was over three miles away. Giving the public a bit more protection was clearly a good thing; but there was some behind-the-scenes muttering by many people in aviation that the ebullient barnstorming approach of pilots who had learned to fly in the daredevil thirties and had emerged victorious from the combat sorties of the Second World War was perhaps not best suited to the new Jet Age. Experimental and even production aircraft were now flying at speeds scarcely imagined a decade earlier. Such performance imposed enormous stresses on airframes, to say nothing of the men who flew them. In

Britain, at least, not even pressurised cockpits were standard, let alone *g*-suits to protect against pilot black-out or red-out in high-*g* manoeuvres. Yet, for all that a new generation of scientifically literate, methodical test pilots like John Derry was beginning to emerge, the participants at Farnborough in 1952 were still acutely aware of the pressures urging them to show off rather than to play it safe. Bill Waterton recorded one pilot's wife saying: 'I do hope my husband makes a better bang than so-and-so. It can mean so much to us.' He later went on to complain in print: 'The ballyhoo has reached crazy proportions. Rivalry between companies has led to them urging their flyers to try to outdo the pilots of rival companies . . . Farnborough, in fact, is no longer aimed at demonstrating an aeroplane's best features to a select, knowledgeable audience, but is . . . something of a Roman holiday.'[12]

John Derry, too, would have been equally aware of these pressures and equally keen to resist them. His former boss had been Jeffrey Quill, the pilot who would be known for having test-flown every mark of Spitfire. 'What is unforgivable,' Quill had once remarked, 'and was at one time all too prevalent, is if pilots use public or semi-public occasions and highly important aeroplanes to try to demonstrate their own daring. I saw too many people roll or loop themselves into the ground, and still more very nearly do it, and I became bitterly hostile towards the slightest lack of professionalism in demonstration flying.'[13] Quill had singled out John Derry as a gifted young pilot who shared his ideals, and Derry's demonstrations all that week in the DH.110 had shown he was never less than professional, always performing within the limits he believed his aircraft to have.

Even so, on the day before their deaths a wire recorder in the cockpit of Derry and Richards's black 110 registered a brief conversation at altitude during their Farnborough demonstration when Derry, experiencing some buffeting, had loosened his straps enough to be able to peer back through the canopy at the swept-back port wing – clearly the first time he had done so:

D: God strewth! I am glad I haven't looked at the wings before at high speed, Tony.
R [looking at the starboard wing from his porthole]: Hell, they're all over the place, waggling about.
D: Waggling about like an ornithopter.
R: Christ.
D: Never mind, not to worry.[14]

Yet not worrying was arguably to cost them their lives within twenty-four hours. It is amazing by present-day standards what risks cautious pilots were prepared to run even when they could physically sense, see and hear that their aircraft's construction was dubious. John Wilson, another de Havilland test pilot, recorded how much he had enjoyed demonstrating the prototype Comet airliner at Farnborough in 1949:

Our fly-past air speed at Farnborough was 340 knots [391 mph], very fast for an aircraft which had only been flying for three weeks. At that time we had not explored in great detail the handling characteristics at higher speeds. We had Chris Beaumont (Chief Test Pilot of the D. H. Engine Company) with us, and he stood at the entrance to the cockpit – every time we pulled 2½–3g to go round the corner Chris found the floor on which he was standing was bulging up. And there was a loud bang at that speed from the nose of the aircraft where the skin panted [flexed] so when we heard this bang we knew without checking the air-speed indicator that we were doing 340 knots. In later years we realised that these were indications of how flimsy the structure really was.[15]

'In later years' presumably refers to 1954, when disasters involving the Comet led to the biggest air-accident investigation ever launched and, arguably, to a downward turning-point in British civil aviation.

Britain really was a different country back in 1952, inhabited by subtly different people in a markedly different economic landscape. Most unfamiliar to modern Britons would be the idea of a flourishing and glamorous aircraft industry that justifiably claimed to share world leadership with only the USA and the USSR, and which in certain respects was ahead of both. It was huge, well funded, and full of frequently trumpeted plans for a world-beating future. Social attitudes, too, were differently premised. To Britons in the twenty-first century who have grown up in an increasingly litigious society and in the protective arms of the Health and Safety Executive, it might seem that many things about Britain some sixty years ago were strange, even quite brutal. It did not occur to the relatives of the victims of John Derry's crash to sue either the SBAC for unnecessarily endangering spectators' lives, or de Havilland for demonstrating a dangerously unfit aircraft. In those days one could often encounter notices pointing out that if you wished to endanger yourself in a hazardous pursuit or place, you did so at your own risk – in exactly the way you climbed trees, fell off bicycles and experimented with fireworks as a child. It would have seemed downright unpatriotic for one of those injured at Farnborough in 1952 to have attempted to make somebody pay for the accident. Risk-taking was a normal part of life: everyone knew that. Had we not just survived a world war?

So back to Farnborough people went the next day in ever-greater numbers, even as the boffins set about painstakingly sorting and labelling the sad wreckage of Derry and Richards's silver beast, which had been collected and piled in a hangar well away from the public area.

Bill Waterton and the World
Air-Speed Record

Derry and Richards's Farnborough crash made a deep impression on me at the time. I was nearly eleven years old and obsessed with anything to do with aviation and plane-spotting. My fascination with aircraft was obviously rooted in the war. I was born in 1941, the year after the Battle of Britain, within a mile of Bentley Priory in Stanmore (then Middlesex, now Greater London), Headquarters of RAF Fighter Command from where the Battle and all Britain's air defence were directed. The skies of my infancy were full of portents and aircraft. In those days one was never very far from an airfield, and my mother used to tell a story of my being so terrified one day by the sudden roar above our garden of a low-flying aircraft coming in to land that I bounced myself hysterically out of my pram into some rose bushes. That was her account, fossilised in the way of such things into the holy writ of a family tale. But I have long wondered whether in fact I might have been excited rather than terrified: impatient to leave my pram in order to see where the aircraft went behind the hedge.

Growing up in post-war Britain there were constant reminders of the scope and inventiveness of our immense aviation industry. By the early fifties our perception as schoolboys was that Britain and aircraft were indissolubly linked, much as the French were with garlic and the Italians with opera. Hadn't the Battle of Britain a mere ten or twelve years ago marked the first time in history that a purely aerial engagement had saved a nation and helped decide the outcome of a war? Certainly my grandmother always told me that whoever controlled the airspace above a country controlled that country, too (a comparatively recent shift in the strategy of warfare). At infrequent intervals in our childhood my sister and I were taken up to town to visit this lady: our mother's mother, an invalid badly crippled by rheumatoid arthritis. This was immediately after the war, when London was grimy and pockmarked with bomb sites where groundsel and buddleia were already well established. On the walls at street level were fading arrows pointing to the nearest air-raid shelters. Granny Sal was transplanted through a grim series of nursing homes in the Earls Court area, the last of which was demolished decades later when the Cromwell Road was drastically widened. She always sat in a high-backed chair, swathed in shawls, one foot on a footstool, the small table in front of her piled with books and newspapers as well as the clock from a Spitfire that I coveted.

A woman of acute intelligence and quickness of mind, she bore her enforced and painful immobility with great stoicism. She had worked in the Censorship in the First World War and proved disappointingly loyal to the Official Secrets Act she had signed, referring tantalisingly to things that were still too 'hush-hush' to talk about. She maintained a razor-sharp interest in politics, as befitted the widow of Donald im Thurn, the sometime MI5 agent who had played a none-too-creditable role in the Zinoviev Letter affair, a conspiracy hatched by White Russians and high Tories to scare the British electorate with the communist bogeyman in the run-up to the 1924 general election, which the first Labour government duly lost. Like my grandfather Donald, Granny Sal was a firm Tory, a devout patriot, an unquestioning admirer of Winston Churchill and deeply suspicious of American motives. Like most Britons who had lived through the Battle of Britain she unhesitatingly put the Royal Air Force jointly with Churchill at the top of her list of the nation's saviours. She had an extensive collection of Stationery Office publications about the RAF as well as numerous cuttings of poems by airmen that had found their way into newspapers.

I have an abiding memory of one visit to Granny Sal when she read a poem to us. Although I remembered nothing else about it, one particular phrase stuck with me for the rest of my life: *'the conquer'd air'*. Later, at university, I tried half-heartedly to track it down but without success. It wasn't until our mother died and my sister and I inherited her books that I found it. It came in some lines that Granny Sal had copied onto the flyleaf of a wartime Faber collection, *Verses of a Fighter Pilot* by Flying Officer A. N. C. Weir, DFC, who was killed in action in 1940. It went:

> The time will come, when thou shalt lift thine eyes
> To watch a long-drawn battle in the skies,
> While aged peasants, too amazed for words,
> Stare at the flying fleets of wondrous birds.

> England, so long the mistress of the sea,
> Where winds and waves confess her sov'reignty,
> Her ancient triumphs yet on high shall bear,
> And reign, the sovereign of the conquer'd air.

A note at the bottom that her distorted hand with its bunched knuckles had carefully made with her silver Waterman pen correctly identified this as the free translation of a passage from Thomas Gray's Latin poem *Luna habitabilis*, written in 1737.

I can see exactly why these lines appealed to her, both as remarkably visionary and as predicting a future in which England would maintain its superiority in the air. *Was it not foreseen?* In the late nineteen-forties this would perhaps not have appeared an outlandish prophecy. Hardly a day passed without some announcement that a British aircraft had broken yet another record – whether of speed or altitude or distance – and my young mind absorbed like a sponge my grandmother's fervent belief in Britain's enduring greatness and its dependence on an unrivalled prowess in aviation. High above her Blitz-cracked ceiling the skies were full of wailing Gloster Meteors and de Havilland Vampires, as well as the streaking prototypes of what would hopefully become the next generation of world-beating (and export-achieving) aircraft. No matter what disillusionments the next two decades were to bring, and no matter how soon Gray's fond eighteenth-century vision was to be betrayed by twentieth-century political reality, my grandmother's emotional reading of his lines had an indelible effect on me. Just as the zoologist Konrad Lorenz induced his motherless baby geese to imprint on him, so my crippled grandmother imprinted on me her private version of Britain: a brave new Jet Age in which we were sovereigns of the conquer'd air.

As Britain's wartime saviour and home-based defender the RAF was not like the Royal Navy, whose warships could be sent far around the world and had anyway proved vulnerable to attack from the air, endowing them with a faint but distinct aura of obsolescence. Once the army had been rescued from Dunkirk in 1940, the last fall-back of the nation's defence was the RAF's Fighter Command. Despite great losses it had proved heroically up to the task of preventing the invasion of England's green and pleasant land by armed foreigners intent on polluting it with Nazism. As a boy, therefore, I always associated it quite specifically with the gallant defence of the realm: a friendly, domestic sort of guardian enabling us all to sleep soundly at night. I think most of my school friends imagined that RAF pilots were pretty much exclusively 'chaps like us': white, middle-class, ex-public-school boys. My grandmother, though, always emphasised the RAF's sheer heterogeneity: that its pilots were black and white and brown, Indians and South Africans and Australians and Poles and Free French and, yes, even Americans. According to her they represented a spontaneous closing of civilised ranks everywhere against Nazi barbarism. Where better to make their heroic stand than in the last European outpost of freedom, the Empire's motherland? It was some years before I discovered that the test pilot I most admired as a boy was not English at all, but Canadian.

The names of one's childhood heroes have a habit of

ringing on faintly in the background of adult life, gradually acquiring a mythic quality so that when heard again they can trigger a burst of affectionate nostalgia as for a vanished brand or a row of cottages long since laid waste by the wrecker's ball. That former household names are unknown to later generations is banal enough in itself; but there is the sadder implication that their achievements themselves are forgotten, as though they have no relevance for a later age. In the case of the test pilots who flew their way into Britain's headlines in those seemingly magical ten years after the Second World War this is not merely unjust but plain wrong. The Jet Age with its heroic challenging of those great unknowns, transonic flight (the speed at which the vagaries of the airflow over certain parts of an airframe make them go supersonic before the aircraft as a whole does) and the sound barrier, is indeed a vanished era. But the pilots who tested those temperamental early jets with such methodical and selfless daring left a legacy that daily benefits every modern air traveller, whether flying first class to New York or cattle class to Ibiza. The sheer safety of flying now is rooted in the deadly precariousness of flying then. For that, if for no other reason, one rejects the facile parallel sometimes drawn between the fame of those post-war pilots and that of contemporary Formula 1 drivers. Britain's post-war test pilots in their cloth helmets and ratty overalls often took home as little as £800 or £1,000 at the end of a year of living with imminent death on a daily basis. Far from being a weekend spectator sport, their daily work built gradually into a new knowledge of aerodynamics that today determines every aircraft design, military or civil.

As boys, we aircraft maniacs back in the fifties were as familiar with the names and careers of newsworthy test pilots as any sports fiends were with their cricketing or soccer heroes, and the daily newspapers accorded them at least as much attention when they achieved a new record or other memorable feat (such as cheating death by inches). In any year our *Eagle* or *Schoolboy's* diaries might well have contained one or more photographs of such aviation heroes as Roland Beamont, John Cunningham, John Derry, Neville Duke, Roland Falk, Mike Lithgow, 'Bill' Pegg or Peter Twiss. We were even aware of some older-generation test pilots such as 'Mutt' Summers and Jeffrey Quill. Most of us had a particular idol. The pilot to whom I accorded my special loyalty was W. A. 'Bill' Waterton, I think because he was chief test pilot at Gloster and held records in the Meteor – an aircraft for which I felt affection because it was so familiar from its daily appearances over our house. In addition, Waterton was famously testing the prototype Javelin, the top secret twin-engined delta interceptor eagerly awaited by the RAF and rival to de Havilland's DH.110. (It was the dicey Javelin that he had taken up at Farnborough in 1952 in the immediate aftermath of Derry and Richards's crash, following Neville Duke's Hunter into the air to demonstrate that the show must go on.) Over the intervening decades my interest in him, politically innocent as it was in my aero-spotting pre-teens, proved to be perceptive. Bill Waterton can now be seen not only as one of the most heroically capable test pilots of the period, but as a remarkable prophet. St Matthew observed with some acerbity that a prophet was not without honour, save in his own country. In his ornery way, Canadian-born Waterton went one better and became a prophet without honour in both his own and his adopted country. His crime was effectively to have foreseen the degradation of the British aviation industry by identifying the canker at its roots, and this at a time when we the public thought it most gloriously flourishing. For this he was eventually sacked and vilified. To judge from the Soviet-style oblivion to which he has been consigned for over fifty years on both sides of the Atlantic, it appears that he still has not been forgiven.

Bill Waterton was born in Edmonton, Alberta, in 1916 and brought up in nearby Camrose, where his father was chief of police. The family had originally come from near Owen Sound in Ontario, where they had farmed in conditions of pioneering hardship and where there is still a Waterton Creek and Waterton headstones to be found in local cemeteries. In later years Bill liked to trace his lineage back through the magnificently eccentric English naturalist and explorer, Charles Waterton (1782–1865): evidence perhaps of his pride in taking an individualist's path as well as of a lifelong Anglophilia. Young Bill's upbringing no doubt made for a character that combined belligerence with probity. It was likely that a police chief's son would be challenged by other boys, and certain that his father instilled in him an ethos of strict self-reliance and a refusal to back down from a fight. It was equally likely that such a lad might go on to Canada's Royal Military College (where he won a gold cup for boxing) and to serve briefly in the 19th Alberta Dragoons, when he appears in a contemporary photograph sitting his horse in full cavalry fig, complete with long sword, an incipient moustache and a somewhat wooden 'They shall not pass!' expression. He had already acquired a taste for flying when at the age of fifteen he and a friend had pooled their pocket money for the two dollars apiece required by a visiting barnstormer, Ernie Kubacek, to take them up in his red Alexander Eaglerock biplane. The former German First World War pilot treated the boys to a series of barrel rolls, loops and stall turns and in Bill's case, at least, the experience must have proved formative. 'You felt free! You saw things from the air that you never saw on the ground. It was also like riding: there was this *physical* aspect to it.'[1] In 1938, at the age of twenty-two and responding to the likelihood of war in Europe, he applied for a commission in both the Canadian army and the Royal Canadian Air Force, though without eliciting a positive enough response. At this time the RAF, also foreseeing war and a shortfall of pilots, was recruiting from all over the Empire by advertising short-service commissions in local newspapers. Bill answered one of these, duly passed a medical examination and an interview and in March 1939 was one of eight Canadian RAF applicants who sailed for Liverpool aboard Canadian Pacific's *Duchess of Richmond*.

For their basic training Waterton and his colleagues were first sent to No. 5 EFTS (Elementary Flying Training School) at Hanworth in Middlesex. This was a civilian school operated by Flying Training Ltd, a company run by Blackburn Aircraft Ltd using their own B-2 biplanes, similar to the de Havilland Tiger Moth. Bill achieved high marks and after fifty-one hours' flying went on as an acting (probationary) pilot officer to RAF Uxbridge for full basic training. This he finished with a posting to Gloucestershire, for whose countryside he was always to express great fondness and which became the setting for his future career as a test pilot. On the outbreak of war and with 144 hours in his log book he was sent to 242 Squadron, an all-Canadian outfit at Church Fenton, not far from York, where he signed up for a six-year short-service commission. After some months the squadron was sent some Hawker Hurricane Mk 1s and it was as a Hurricane pilot that Waterton was posted south to Kent. In the build-up to the Battle of Britain he began flying combat sorties from Biggin Hill and Manston. One day late in May 1940, high above France while providing air cover for the Dunkirk evacuation in Hurricane L1852, he 'felt a bump' and passed out. He was seen spiralling down but managed to regain consciousness intermittently, somehow setting a course back across the Channel where he crash-landed. His terse log book entry reads: 'Hit – out of control near Dunkirk at 19,000 ft. Engine bad vibration. Crash-landed near Dover – concusion [sic] – head injuries. Oxygen loss.' Bill was taken to the Royal Masonic Hospital in London with back and head injuries that took him off the active list for nearly three months.

In mid-August 1940 he returned to 242 Squadron, which had meanwhile sustained heavy losses among his Canadian comrades. It was now based at Coltishall in Norfolk under the irritable command of Douglas Bader, the celebrated pilot who flew with artificial legs. Bader promptly sent him to No. 6 OTU (Operational Training Unit) to get his hand back in: a move which probably determined the future course of Bill Waterton's flying career. For he turned out to be an exceptionally good flying instructor – perhaps surprisingly, given his forthright nature. But his former skill as a champion boxer had taught him how to master his emotions, and where flying was concerned he was invariably patient and methodical. The Battle of Britain was then nearing its climax and the RAF's critical shortage was less of aircraft than of properly trained pilots. A man of Bill Waterton's skills was rated more valuable as an instructor than as a combat pilot, and he was destined to spend the next two years instructing, both in the UK and back in Canada (where he spent the whole of 1942). In wartime you go where you're sent. Being a flying instructor is notably less glamorous than being a fighter pilot, and there must have been times when he chafed at the lack of action. On the other hand it is rewarding to do something you're good at, and to be recognised for it. While in Canada his wing commander wrote in Waterton's log book: 'You are setting a most excellent example and doing a first class job of work for which I am very grateful. Keep it up.' His squadron leader added a schoolmasterly note: 'An above-average instructor who gives excellent demonstrations. Aerobatics are very smooth. Instrument flying requires practice.' Having successfully taught 509 pupils without a single serious accident, let alone a fatality, Waterton was awarded the first of his two Air Force Crosses.

There must indeed have been something exceptional about his flying. Once back in England in 1943 he was assessed 'Above the average' (a rare accolade in the nit-picking RAF) and was returned to Fighter Command.

Here he test-flew a variety of experimentally equipped Spitfires at extreme altitudes and carried out meteorological sorties high above occupied Europe during which he was required to make a careful record of instrument readings. In May 1944 he was sent to AFDU (Air Fighting Development Unit) at RAF Wittering near Peterborough. This was a coveted posting because the unit was known to be reserved for the very best pilots. It was also a combat unit where you couldn't be shot down. Fighter Command had set up the AFDU to evaluate captured enemy aircraft, putting them up in mock combat against every variety of Allied aircraft. High above Lincolnshire Messerschmitts and Focke-Wulfs and Junkers 88s, all wearing RAF roundels, fought it out against an assortment of Hurricanes, Spitfires and Typhoons, using gun cameras in place of live ammunition. From then until the end of the war Waterton flew sundry types of British, German and American aircraft. Apart from being wonderful flying experience it produced vital information. Methodical testing revealed the strengths and weaknesses of the different types and as a result armaments, propellers, fuel systems, instrumentation and airframes were modified accordingly, as well as fighting techniques and tactics (*never* try to out-dive a Focke-Wulf 190 and *never* try to reverse a turn when one is on your tail).

The war in Europe ended in May 1945 and in August Waterton, now a squadron leader, flew a jet aircraft for the first time: an underpowered Meteor I. Three months later he flew the more advanced Meteor III and realised that the day of the piston-engined fighter was finally over. It is doubtful that he could have suspected his own peacetime career was about to take off. The ending of the war had thrown everything into confusion, especially for a serviceman whose short-term commission was nearing expiry. The RAF was suddenly characterised by an atmosphere of demoralisation. As Waterton wrote later: 'The vast wartime expansion of the RAF seemed to have left it without any overall peacetime policy. No one knew where he stood. Those like myself who hoped for permanent commissions were disheartened to see themselves overlooked in favour of younger, end-of-the-war recruits. Discipline was dreadfully lax...'[2] Someone might have tipped him the wink that it would be worth his while to hang on for a bit, however, for in spring 1946 came the news that he had been selected to fly a Meteor in the High Speed Flight unit based at Tangmere, Sussex.

By the end of the war it was becoming clear to pilots everywhere that the new jet engine represented the future. The need for speed was paramount in interceptors hoping to meet incoming bombers before they could reach their targets, and there were limits – both theoretical and practical – to how fast a piston-engined aircraft could fly. Beyond a certain speed the propeller's whirling disc caused more drag than propulsion. Nor could that be overcome by making an engine powerful enough to turn the airscrew still faster. At a certain point the tips of the blades began to go supersonic, greatly reducing their efficiency as well as setting up potentially disastrous vibrations. In fact, anyone like Waterton who had flown the faster fighters in dives – Spitfires, Focke-Wulf 190s, Mustangs and so on – had long discovered that when the aircraft as a whole began approaching supersonic speeds strange and frequently fatal things could happen.

In the opening sequence of David Lean's 1952 film *The Sound Barrier* a lone Spitfire performs joyful aerobatics in the sky above the white cliffs of Britain's south coast. It must be late in the war because the pilot is paying no attention to the possibility of being surprised by enemy aircraft, while the anti-aircraft gun crews down below are lounging at their ease on the grass, idly watching the tiny aircraft high above them. We join the pilot in his cockpit as, in an access of youthful exuberance, he noses the aircraft over into a steep power dive as though to see how fast it might go. After a few seconds of rapidly increasing speed with the engine note rising to a wail, the cockpit and wings suddenly begin to judder and shake violently and there are snatched glimpses between the clouds below of the ground coming closer. In a panic the pilot hauls back on the stick with both hands, and such is the physical strength required to bring the aircraft out of its dive that he ends up slumped deep in his seat before he can level out and skim low over the fields.

The pilot was played by John Justin, who had actually been a Spitfire test pilot and flying instructor during the war. Back on base after this sequence, reminiscing to his squadron buddy (played by Nigel Patrick), Justin muses on the difficulties he has just encountered. 'The harder I pulled, the more the nose went down. It felt for a moment as if the controls were reversed ... There was a lot of buffeting, too. It was almost as if I'd run into a ... a solid sheet of water or something.'

This sequence perfectly fixes the sound barrier's mythic

status in the decade following the war. The episode was almost certainly based on a famous incident in 1944 when Squadron Leader Tony Martindale had dived a Spitfire XI from 40,500 ft to 27,000 ft and reached Mach 0.91 (equivalent to 620 mph at that altitude). The dive was part of a Farnborough research project and the aircraft was very accurately instrumented. The pilot blacked out under the 11g recovery, the propeller and part of the engine were torn off and the wings deformed. Yet the aircraft did not break up and Martindale recovered consciousness in time to glide it in for a dead-stick landing (i.e. one made with no power): a tribute to both the Spitfire and its pilot. Calculations later showed that the massive Rolls-Royce Merlin engine had contributed a mere one per cent to his speed: virtually all the rest was down to simple gravity. It was this dive that proved the theoreticians correct: that the drag of its propeller would always prevent any piston-engined aircraft from going supersonic.

The gremlins that lay in wait for the high-speed pilot – the intense shaking, the unexpected nose-down tendency and the flutter of control surfaces (ailerons, elevators, rudder, etc.) that could lead to their complete disintegration within seconds – were well known by then and had been experienced by scores of pilots flying high-performance fighters. A great many of those did not survive to sit on a friend's bed and muse about what had happened. Their aircraft were either seen to disintegrate in mid-air in a glittering cloud of aluminium alloy or else they never pulled out of their dive. There was nothing to be learned from a deep crater in a field after an aircraft had slammed into it at 500 mph; searchers did well if they could identify a small rag of scalp. Typical of this sort of accident was the one that befell Squadron Leader E. B. Gale, a Canadian test pilot with RAE Farnborough, on 25 May 1946. He took off in a North American Mustang Mk III intending to dive from 40,000 ft in order to provide data that would reveal how various parts of the wings behaved when nearing the speed of sound. 'The aircraft was seen diving vertically at extremely high speed before impacting a crop field. Most of the wreckage was so deeply embedded as to be irrecoverable.'[3]

The phenomenon that pilots like Squadron Leader Gale encountered was known to aeronautical science at the time as 'compressibility'. The buffeting came about because at a certain speed – and it is a different speed for each aircraft design and even for individual aircraft – the flow of air over its wings changes from being smooth or 'laminar', hugging the surfaces, to becoming turbulent and breaking away. It was recognised that this process was related to the speed of sound in air. Until a certain speed is reached the airflow can find its way around an airframe's various curved surfaces; but beyond that speed it no longer has time to do so and instead builds up in front of the surfaces in the form of a shock wave, causing sudden drag that can have disastrous effects on the aircraft's handling. Compressibility effects on airflow had been known about since at least the mid-thirties but perhaps were not well understood by enough people outside Germany, where the phenomenon had been intensively studied. In 1935 a German aeronautics professor named Adolf Busemann had given an open lecture at one of the international Alessandro Volta science conferences in Rome. One topic that year was high velocities in aviation and Busemann read a paper about supersonic airflows, his data based on experiments he had been carrying out in Germany in the world's only supersonic wind tunnel. In this landmark lecture he proposed a swept-wing design for high-speed flight since he had discovered that sweepback was a way of delaying the effects of compressibility.

Despite his paper being heard by international scientists, including some from Britain and the United States, it was largely treated as being of merely academic interest. In 1935 few aircraft in general service were capable of 300 mph, let alone 500–600 mph. Even so, the last Italian and British Schneider Trophy aircraft – racing floatplanes with enormous supercharged engines – had exceeded 400 mph and such machines were clearly forerunners of a much faster generation of aircraft still on the designers' drawing boards. The upshot of the Volta conference was that Busemann went home somewhat unsung, his work having aroused respect but little excitement abroad. During the war he became director of the secret Hermann Goering Aeronautical Research Institute in Volkenröde, near Brunswick. This was something akin to a German version of RAE Farnborough but with vastly superior wind-tunnel and other facilities. There was nothing string-and-sealing-wax about Volkenröde; and using its several supersonic wind tunnels throughout the war German scientists like Busemann and Alexander Lippisch were able to experiment at leisure with different wing designs: swept-back, swept-forward, thin, thick, delta, diamond and even asymmetric. Just how advanced this research was can be seen by the way many of these shapes were destined to be hailed decades later as the very latest British or American top-secret designs, including those for modern 'stealth' aircraft and NASA's asymmetric 'slew-winged' Ames AD-1.

The line usually adopted by historians is that no Allied aircraft designers took the effects of compressibility seriously into account until at least 1945, presumably since almost no one was envisaging speeds above 500 mph and consequently didn't bother with questions of sweepback. Yet this is simply not borne out by the facts. To show that in Britain even 'the powers that be' (in the ubiquitous biblical phrase of the times) kept abreast of modern aerodynamics, in 1942 the Ministry of Supply asked Armstrong Whitworth's chief designer to produce a wing that the National Physical Laboratory could test for its laminar flow characteristics. Inspired by the results, Armstrong Whitworth projected a swept-back tailless 'flying wing' bomber with four Metro-Vick (Metropolitan-Vickers) turbojets, and a third-scale glider version was built to test the idea. Nothing came of it during the war but it led directly to the company's twin-jet A.W.52 'Flying Wing' that first flew in late 1947. The truth is that aircraft designers everywhere had been toying with just such revolutionary wing designs almost since the beginnings of aviation itself; but in Britain during the Second World War the absolute priority was to produce overwhelming numbers of conventional aircraft. It therefore remains ironic that Germany's advanced aerodynamic research, which encompassed the world's first jet aircraft as well as rocketry, was largely spurred on by the necessity of finding a defence against the huge waves of conventional Allied bombers that so greatly outnumbered the Germans' own air forces. German strategists soon realised that their only hope lay in being able to fly faster than the Allied fighters escorting the bombers, and this in turn encouraged them to design aircraft capable of speeds as high as 600 mph and even faster (as in the case of the jet-powered Messerschmitt 262).

John Justin's late-model Spitfire performing aerobatics above Britain's south coast at the end of the Second World War can now be seen as celebratory of an era passing. Aesthetically beautiful and iconic though they might have been, such aircraft were by then dead tech. The RAF felt this acutely. From now on the future lay with the jet engine and radical wing shapes that would open up a completely new world of flying. That is why there is nothing elegiac about *The Sound Barrier*'s opening sequence. Indeed, Malcolm Arnold's score – a confident C-major march with unmistakable Elgarian overtones – is clearly meant to announce the beginning of a glorious new chapter in British aviation. This is confirmed immediately after the titles by a list crediting the aircraft that would be starring in the film: the de Havilland Comet, the Vickers-Supermarine Attacker, the de Havilland Vampire 113 and the Vickers-Supermarine Swift ('with Rolls-Royce "Avon" engine'). At this level it is a shameless plug for the nation's aircraft industry and must have gone down very well with the Central Office of Information.

At the end of the war Britain had a head start in the Jet Age. With one fighter (the Gloster Meteor) already in service and a second (the de Havilland Vampire) about to become operational, it might be imagined this would have sparked an enthusiastic upswing in RAF morale, with the promise of new aircraft and heady new capabilities. This was no doubt true for the newest intake of young pilots. Yet contemporary evidence, including Bill Waterton's own account, suggests that the passing of the old wartime order was also too much bound up with deep and inchoate emotions, as well as political misgivings, for the service as a whole to give way overnight to a simplistic optimism. Aviators tend to be a conservative lot anyway, and the RAF appeared to have limited faith in the new political order. It has been shown that Battle of Britain pilots overwhelmingly voted Conservative in the 1945 general election. John Collins, the later Canon Collins who co-founded CND, was an RAF chaplain during the war and was radicalised by the experience. 'He recalled the undisguised horror he encountered in many officers' messes following the news that Labour had won a landslide victory: "Was it for this, they seemed to be saying to themselves, that the war had been fought and won?"'[4]

It was inevitable that once the emergency of war was over the RAF would have to move decisively to adapt itself to new technologies and a new world order. It could not for ever rest on its special status as laurelled defender of the British Isles. To combat the demoralisation that was affecting the service and to show the world that Britain was fully in command of the new jet technology, it was decided in 1946 to launch a fresh attempt on the world air-speed record with the Meteor 4. The High Speed Flight had been formed at Tangmere the previous year for Group Captain 'Willie' Wilson's successful attempt and it was clearly felt that national prestige was at stake. Although Britain already held the record since Wilson had raised it to 606 mph in a Meteor 4, the 'powers that be' must have calculated that an exhausted nation – not to mention an RAF suffering from post-war depression – could do with a further boost to morale. Besides, it must have seemed prudent to capitalise on Britain's lead in jet technology while the going was good. The Americans were known to be making excellent progress with their own jet-powered Lockheed P-80 Shooting Star and might soon make an attempt of their own. The High Speed Flight was commanded by Group Captain Edward 'Teddy' Donaldson, himself an exceptional and bemedalled pilot who had commanded 151 Squadron in France from the beginning of the war and then at Tangmere during the Battle of Britain. He scored several victories and was twice shot down. The two other pilots were Bill Waterton and Neville Duke, then a flight lieutenant with a remarkable wartime combat record. For Donaldson to have chosen Waterton as his number two in this elite group meant that Bill must have acquired unimpeachable credibility and be very highly thought of in all the right places – for after the wartime glamorisation accorded it, the RAF was a service in which envious backbiting had lately reached the status of an art form. It might also have helped that Donaldson had spent three years at McGill University in Montreal and had a soft spot for Canadians, although he would never have let that outweigh matters of technical competence.[5]

For his part Waterton considered the publicity surrounding the new record attempt premature and ill advised. At the time he probably kept this contrary thought to himself, even though such tact would have been quite uncharacteristic of him. Some ten years later he recalled: 'In an understandable effort to bolster Britain's post-war prestige, the press had built the High Speed Flight into something approaching the super-colossal – hyperbole encouraged by official quarters which should have known better. My own feelings were to do it first and trumpet afterwards.'[6] In fact, stimulated by the resoundingly patriotic media coverage, thousands of people poured into Bognor Regis each day to watch the three Meteors practising. In order to qualify under FIA (Fédération Internationale Aéronautique) rules, the course was three kilometres long; it lay just offshore between Rustington and Angmering: in jet aircraft terms, just down the road from Tangmere. The timed runs had to be made below seventy-five metres (246 ft), two in each direction so as to negate any effects of wind. In fact, though, the Meteors were flying as low as eighty feet above the sea at over 600 mph, which provided the enthusiastic crowds with the additional thrill of knowing that the least miscalculation or mechanical failure would give them ringside seats for a spectacular disaster.

Those practice sessions in July 1946 were probably the greatest crowd-pleasers of their kind since the pre-war days of the Schneider Trophy races. Even so, dissenting voices wrote to the papers. 'We didn't mind the noise when the RAF was fighting the Luftwaffe,' one local resident observed, 'but the scream of jet planes flying at 250 feet is unbearable.' From the depths of austerity Britain others quite reasonably wondered, 'Is this speed bid worth while? What good is it going to do anybody even if the record IS put up to 625 mph or more?' And at least one newspaper reported shameless profiteering: 'Visitors to the "grand-stand coast" are being exploited! Twenty-eight guineas a week for bed and breakfast only [some £900 at today's prices] was demanded from a couple who, with their two children, were seeking rooms to ensure they saw the speed bid.'[7] Suddenly the question of Bill Waterton's nationality arose when he was interviewed, despite his flying a British aircraft. 'Even if S/L W. Waterton, one of the High Speed Flight, is the pilot to break the record, it will still be held by Great Britain. He said in a strong Canadian accent: "I am in the RAF and I am doing it for the RAF. I am a British subject according to my passport, born of Irish parentage and resident in Canada."'[8] It seems incredible today that a decorated and hand-picked pilot from one of the Dominions who had come to the UK expressly to fight for Britain should have had his eligibility questioned on grounds of nationality, and not least that a Canadian born and bred should have had to hedge matters further by describing himself as merely 'resident' in Canada, as though for his first twenty-two years he had just been passing through the land of his birth. Yet as Martin Francis has recently documented, despite something like forty-six per cent of all the RAF's aircrew having come from Australia, Canada, New Zealand or South Africa, discrimination and tensions between them and home-grown RAF personnel were often rife, even if less overtly racist than in the case of black or Indian aircrew.[9] Not surprisingly, Bill Waterton felt the implied slur acutely. By this stage he had taken on a considerable degree of protective colouring, having long sported an RAF-style handlebar moustache and modified his Canadian accent to the point where, far from being 'strong' as the newspaper hack reported, it was often barely detectable.

The Gloster Meteor was only Britain's second jet aircraft. The first, the experimental Gloster Whittle E.28/39, had flown in May 1941: a simple, single-jet design built to test Frank Whittle's pioneering new jet engine, which had taken the little aircraft to a respectable 466 mph. The Meteor was much more ambitious and planned from the first as a fighter. It was a straight-winged design with an engine on each wing and with the tailplane halfway up the fin. The twin improved jets gave it a much better turn of speed than its predecessor. The three High Speed Flight pilots' initial task was to determine whose Meteor was the fastest. That aircraft was duly commandeered by Group Captain Donaldson who, as the commanding officer, had first choice. Immediately, severe problems with the engines of all three aircraft were encountered, caused by the special Rolls-Royce Derwent Vs inhaling their own rivets which shattered the turbine blades. One engine also inhaled an incautious Donaldson as he was poking around his aircraft on the ground at Tangmere. As the Group Captain was disappearing into the starboard intake he was saved by a quick-thinking engineer officer, Squadron Leader George Porter, who rugger-tackled him as the engine was shut down. Donaldson was lucky. Unlike many before and after him he survived with nothing worse than heavy bruising and a temporarily off-the-scale heart rate. In August two brand-new, specially finished Meteors arrived from Gloster and Donaldson announced that he and only one other pilot would try for the record. For whatever reason, he picked Bill Waterton over Neville Duke, to Duke's evident disappointment. And then at the last moment, as the pilots were waiting for the weather to clear, the Royal Aero Club – whose officials had to be present in order for a record to be ratified – resurrected the Blimpish issue of whether a colonial could compete in a British team. 'With all civility,' Waterton implausibly wrote later, 'I pointed out that I regarded myself as British, a fact confirmed by my passport; and that if it had been good enough for a Canadian, Lord Beaverbrook, to get the UK its warplanes [he had been Minister of Supply, 1941–2], then it was good enough for me to fly the damn things.'[10] A very Watertonian piece of repartee.

On the day of the attempt his Meteor was still flying port wing low as a result of aileron trouble that Gloster claimed had been corrected. On his first run he found himself in serious trouble from the onset of compressibility at about 580 mph. The aircraft became barely controllable with a mere eighty feet of altitude in hand. Waterton found he needed all his considerable strength just to hold it level and avoid hitting Gertie and Ermintrude, two anti-aircraft balloons tethered to mark the course. At 600 mph 'I stood on the starboard rudder pedal and pressed with all my fourteen stones ... We skidded past the balloons and shot out to sea in a slight climb. Speed fell, and I was able to take one hand from the stick.' He had only narrowly averted disaster but couldn't call Tangmere or Donaldson to explain the problem because in order to achieve maximum streamlining of his aircraft the radio antennae had been removed and the holes blanked off. He flew back to the start of the course again and this time found that if he jammed his left forearm like a rod between the side of the cockpit and the control stick he could just hold the Meteor level, but at a cost. 'At 605 mph it seemed as though every bone from the tip of my elbow to the palm of my hand was in the grip of a giant, remorseless nutcracker: this in addition to the spine-jarring bounce of the aircraft.' He continued in this fashion for three more agonising runs and returned to base disconsolate, his mood made worse by an aching arm and the scepticism he met which was based on Gloster's fiction that his Meteor was 'in tip-top condition'. Next day the Press was exuberant: Donaldson had bumped up the record to 616 mph; Waterton had achieved 614 mph. Later, Neville Duke made an attempt in Bill's Meteor but at first was unable even to get round the course in it. 'This seemed to confirm that there was something wrong with the aileron and not with the piloting,' was Waterton's pointed observation.

In any case Britain had got a new world speed record and the High Speed Flight's existence had been justified. Still, when all the ballyhoo was over the new record was only 10 mph faster than the previous one. The FIA's system of categories also concealed the unwelcome fact that Donaldson's best speed was a full 86 mph slower than the 702 mph that the German Heine Dittmar had achieved in 1944 in a rocket-powered aircraft. And in case one scorned rocket propulsion there was always Heinz Herlitzius's 624 mph, also achieved in 1944 but in a Messerschmitt Me.262, as conventional a jet as the Meteor. Evidently records set by a defeated enemy could be blithely discounted. Donaldson's official world speed record stood for a year until an American Douglas Skystreak achieved 640.67 mph over Muroc in California. The whole enterprise was anyway something of a vainglorious nationalistic charade, although of course the hidden agenda could be summed up in the single word 'sales'. So, having achieved its purpose in 1946, the High Speed Flight at Tangmere was disbanded. The secondment had served Bill Waterton well: his stock was high and he had achieved public recognition even before the news came that all three pilots had been awarded the AFC for their efforts, which in Bill's case meant a bar to his existing medal. On 4 October 1946 his log book records 'Teddy' Donaldson's friendly assessment of him as '<u>Exceptional</u>. A very reliable and most accurate pilot. Has done invaluable experimental work on compressibility problems at speeds in excess of 600 mph.'

One might have thought that after a headmasterly encomium like this, and with the newspapers still buzzing with his achievement, the ambition of Squadron Leader Bill Waterton, AFC and bar, to make the RAF his profession would have been easily satisfied. It wasn't. The RAF refused to give him the permanent commission he sought, fobbing him off with offers of another short-term commission while they 'gave the matter their fullest consideration'. In fact, as he had already surmised, they were giving their fullest consideration to younger men with only a fraction of his experience. To have flown in the war was already to carry the taint of a veteran, which evidently sat uneasily with the 'new' RAF that was taking shape. Like many another pilot in his predicament, Waterton made his own decision. '"If the RAF can't make up its mind after nearly nine years whether or not I'm going to be of use to them," I said, "then it never will. Sorry chaps, but I'm off.'"[11] The makers of the Meteor, the Gloster Aircraft Company, were on the look-out for a test pilot with suitable jet experience and Bill Waterton was the obvious choice. He duly accepted their offer and on 21 October 1946 became a civilian once again. Privately, he was upset at having to leave the RAF, which for all its faults he described as his 'first love'. If he felt it had treated him unfairly, though, he never divulged it. He knew that the service he had known and loved had been shaped – even distorted – by the long emergency of war, and that from now on it would become a very different peacetime entity. It was probably a good moment to be getting out, after all.

3

The Sound Barrier

Throughout the war enormous resources had been poured into Britain's aircraft industry, in line with the sentiments Churchill expressed to the Cabinet and Chiefs of Staff during the Battle of Britain when he said: 'The Navy can lose us the war, but only the Air Force can win it. Therefore our supreme effort must be to gain overwhelming mastery in the air. The Fighters are our salvation, but the Bombers alone provide the means of victory.'[1]

The upshot was that by the end of the war the aircraft industry was Britain's biggest. As it took stock of the industrial giant it had inherited in 1945, the new Labour government identified some twenty-two major companies such as Avro and de Havilland and Handley Page and Gloster, all designing and building airframes, as well as nine independent manufacturers of aero engines who included Bristol, Armstrong Siddeley and Rolls-Royce.[2] All these companies were still more or less on a war footing and geared up for mass production. They were tightly interwoven into a nationwide network of subcontractors supplying them with components. These ranged from the manufacturers of Hispano cannon and radio sets to cottage industries consisting of groups of women sewing marker tapes onto undercarriage locking pins that ground crews needed to visually identify and remove before an aircraft was cleared to fly. With the ending of six years' emergency this network would have to be unravelled. It was obviously the right moment to rationalise the whole of British aviation, scaling things back and amalgamating whole groups of companies, the better to concentrate on researching and developing the new generation of jet aircraft. The same, too, would have to apply to the RAF, with thousands of conscripts being returned to Civvy Street and a leaner professional air force formed for Britain's future defence.

But defence against whom, and what sort of attack? From a victor's point of view an immediate aggressor was hard to envisage, still more to fear, since all plausible enemies were smashed and in ruins. The USSR was still theoretically an ally, even though doubts were increasingly being expressed by anti-communists like the new Labour government's Foreign Secretary, Ernest Bevin. But now British politics repeated themselves. Immediately after the First World War Lloyd George had announced his 'ten-year rule' which declared that no new war was conceivable for at least another ten years. In 1945 Attlee followed suit by deciding that since in theory Britain now had the capability of building an atom bomb of its own there could be no war for at least ten years and therefore no completely new fighters would be needed for either the RAF or the Royal Navy until 1957. Military aircraft already on companies' drawing boards might conceivably be officially backed, but nothing beyond that unless they were purely for research. 'This was a fatal error of judgement,' a well-known aviation commentator was to record; one which was to cost Britain a complete generation

EMPIRE OF THE CLOUDS

of fighters. 'While the United States and the Soviet Union were making tremendous efforts to produce operational swept-wing interceptors, Britain chose to ignore the situation.'[3] Not many years had to pass before first the Korean War and then the Suez invasion did indeed show how wrong Attlee's prediction had been, particularly in the skies over Korea. The RAF could never have put up a fighter comparable in performance to the American F-86 Sabre, let alone to the Russians' MiG-15. For the first time in over ten years Britain's front-line combat aircraft were manifestly inferior to those of the enemy.

Equally fatally, the Attlee government side-stepped the much-needed rationalisation of the huge aviation industry. At a bureaucratic level it scarcely helped its own chances. In October 1945 the old Ministry of Aircraft Production and the wartime Ministry of Supply were merged into a new peacetime Ministry of Supply of truly elephantine proportions, leading to even more sclerotic decision-making than before. (In 1956 *The Aeroplane* was to describe the MoS as 'that monstrous octopus'.[4]) But a further reason for the government's failure to slim down Britain's aviation industry was that many of the famous companies were still in the hands of their founders, a good few of whom (like Geoffrey de Havilland Sr) had themselves been pioneer aviators. These were generally tough and autocratic men who had not the slightest intention of seeing the precious identity of their family business diluted, and certainly not at the behest of a socialist government. On questioning, their loyal workforces mostly proved equally inflexible.

This was hardly surprising, given how comfortably

situated they were. During the war the private companies that made up the aero industry had effectively been nationalised by being brought under the direct control of the Ministry of Aircraft Production. One result was that any competition between them to win the various contracts for new types of aircraft put out to tender by the Ministry came without the peacetime financial penalty for failure. No company that was part of Britain's vital defence could be allowed to go under simply because it had produced a dud aircraft. This was an agreeable position for any industry to be in. The tradition was anyway for new designs to be invited from individual companies on a rotational 'Buggins' turn' system. So although the government now had every reason for slimming the industry down as quickly as possible, the industry itself had every reason for digging its heels in and resisting any policy that might challenge the comfortably subsidised status quo. As their final weapon, most of the companies were household names indelibly associated with the famous aircraft and engines that had helped win victory. No politician was eager to be acclaimed by the newspapers as personally responsible for liquidating any of a dozen hallowed icons. Even so, the process need not necessarily have been traumatic since there was a precedent. In 1935 the Hawker Siddeley Group had been formed when Hawker Aircraft Ltd acquired the Gloster Aircraft Co. and the Armstrong Siddeley Development Co., and this process had not fatally damaged any of the companies involved, which retained a considerable degree of autonomy within the group. Still, with so much else to do by way of getting the country back on its feet the new

Labour government evidently felt this was one confrontation it could do without. The consequences of this failure to act were to prove ruinous.

At this distance it is easy enough to criticise the inability of the wartime bureaucratic machine to adjust to peacetime aviation, and it undoubtedly did betray simultaneous symptoms of frenzy and paralysis. Yet it had a good solid basis to build on because by VE Day in May 1945 British aviation was generally in a very strong position. The country's largest wartime industry had also been the most inventive. When all the requisitioned 'shadow' factories such as Morris at Cowley went back to making civilian motor vehicles again, they began turning out technological fossils in the shape of cars that in mechanical essence were little different from the ones they had last produced in 1939. Not so the military aircraft industry, which had progressed beyond recognition. Having enjoyed liberal funding for so long, the aviation companies were awash with all sorts of ambitious research projects. If ever there was a moment when the industry could reasonably look forward to a period of glorious new achievement that might qualify as 'golden years' this, on the brink of the Jet Age, was surely it.

Being Britain's first production jet aircraft, the Gloster Meteor, the type in which Bill Waterton attempted to break the speed record in 1946, was inevitably a little crude as well as badly underpowered. In the early days it was also quite testing to fly. This was a generic problem with all early jet aircraft, in that their engines were comparatively weak and they were nothing like as responsive to the controls at low speeds as the propeller-aircraft pilots were used to. When a conventional aircraft began its take-off run, the propeller's slipstream passing over the flying surfaces like the rudder and elevators meant that as it picked up speed the controls quickly became 'live', and even when taxiing the aircraft could be steered with some gentle dabs on the rudder pedals. But without this slipstream the early jet aircraft felt very sluggish on the ground by comparison and were steered like bulldozers by braking one or other of the wheels. Nor was there much in the way of acceleration until the gathering speed forced air into the intakes fast enough to give the engine something to 'bite' on and become more efficient. And lastly, the early jet engines were amazingly thirsty. One senior RAF officer allowed to fly the experimental Gloster Whittle E.28/39 remarked that it was the only aircraft he'd ever flown where you could actually see the fuel gauge needle moving while the engine was running. A related problem was the availability of the new fuel. At that time there were few airfields in Britain with a supply of aviation kerosene, and a Lancaster bomber fitted with a large tank of the stuff usually accompanied the Gloster Whittle wherever it went.[5]

The twin-jet Meteor first flew in March 1943 with test pilot Michael Daunt at the controls. Daunt, who had become Gloster's chief test pilot after the E.28's first pilot, Gerry Sayer, was killed in a collision while flying a Hawker Typhoon, had himself only just recovered from six months' convalescence after being severely injured in a crash. No sooner had he recovered than he earned himself the distinction of becoming one of the first people to be sucked into a jet intake on the ground. He was wearing a large leather flying coat which had acted like a sail, and but for the prompt action of the pilot in chopping the throttle Daunt would have been killed. These glorified vacuum cleaners could bite even at their front ends.

It didn't take long for pilots to discover that the early marks of Meteors also bit. New technologies seldom show a complete break with the past. Early motor cars had gone on looking like horseless carriages, with their wooden-spoked wheels and their owners pointlessly exposed to the elements like stage coach drivers, far longer than technological necessity demanded. In the case of an early jet like the Meteor its engines were placed one on either side of the fuselage as though they were conventional motors, far enough out along the wing for their imaginary propellers to have sufficient clearance.

One consequence of this arrangement was asymmetric handling if one of the engines failed, as was fairly frequent in the early days of jets. In effect, this meant that one wing stopped dead while the other slewed viciously around in a flat skid. This was particularly dangerous if it happened at a critical moment, such as when the pilot was turning finals on the way in to land. It then became essential to keep up flying speed and not to stall; but this would normally entail gunning the remaining engine, which inevitably increased the yaw. The pilot had to counteract this potentially fatal swerve with the rudder by means of sheer physical strength. It sometimes took 300 lb of pressure on the pedal to bring the aircraft back into some semblance of a straight line or, as one Meteor pilot put it, the equivalent force needed by a sizeable man doing a full knee bend on one leg.[6] Not surprisingly, this accounted for quite a few casualties in those days before the introduction of power-assisted controls, not to mention ejector seats. Another quick route to disaster was the Meteor's so-called 'phantom dive', which an unwitting pilot could easily trigger merely by forgetting to retract the air brakes before lowering the undercarriage. If the aircraft yawed towards one side while the air brakes were out the Meteor could be abruptly tipped into a dive. It took time and several fatalities to trace the real cause of the problem. It turned out that because the Meteor's main wheels went down one before the other this momentary drag on one side of the aircraft induced a yaw that, though short-lived and slight, could still be enough to stall one wing. Since this commonly happened at low altitude when coming in to land it was often fatal.

Nevertheless, the Meteor was gradually tamed by strong-legged and strong-nerved test pilots into becoming the first Allied operational jet fighter, although by then it was too late in the war to make much difference in combat. But from the summer of 1944 616 Squadron, based at Manston, began using Meteors to tackle V-1 'doodlebugs' as they stuttered across Kent on their way to London. These flying bombs were the world's first cruise missiles, crudely but quite reliably powered by pulse jet. Some were downed by the expedient of a Meteor flying alongside the V-1, slipping a wing tip beneath the bomb's wing and flipping it over far enough to confuse its inertial guidance system so that with luck it would plunge harmlessly into a hop field. The all-metal Meteor was tough enough and fast enough for this procedure to be merely highly risky rather than downright dangerous and was repeated several times with success. In fact, it was probably less hazardous than shooting the V-1s down from the air since if they exploded they could take their attacker with them. Nevertheless, shooting them down was the method favoured by Roland Beamont, the future test pilot of English Electric's Canberra bomber and Lightning fighter, who accounted for thirty-one V-1s in his propeller-driven Hawker Tempest.

As the crowds at Bognor Regis watched the brightly polished Meteors of the High Speed Flight flash past in the sun in the summer of 1946, they must have felt that Britain's future in the leading ranks of aeronautical nations was secure. And yet, all unbeknownst to them, a decision had just been taken that probably damaged irreparably Britain's chances of being the first country to break the sound barrier. It was an extraordinary tale. Back in September 1943, even before the first Gloster Meteors became operational, the Ministry of Aircraft Production with its habitual inscrutability suddenly issued what was arguably the most forward-looking specification ever drawn up by a country in the depths of a major war. It was for an experimental jet aircraft capable of reaching 1,000 mph at an altitude of 36,000 ft.

At that time Britain's aircraft designers were still thinking that, with luck, speeds of up to 600 mph ought eventually to be possible, and here was the Ministry apparently talking airily about 1,000 mph. In his book about the project[7] the late test pilot Eric 'Winkle' Brown, who was to have flown the aircraft, suggested that the supposed goal of 1,000 mph had been triggered by a British intelligence document describing a secret German project for a 1,000 km/h (621 mph) aircraft. This was seen as a challenge that would have to be met – hence the Ministry requirement. However, the lady who typed this out apparently read it as 1,000 *miles* per hour. Yet this theory seems implausible. It is hardly likely that a typo of such consequence would have been overlooked, especially since boffins at Farnborough had already stated that in their view a speed of 1,000 mph was aerodynamically unfeasible.

So MAP's challenge was a severe one. Such speeds were completely unknown territory. Nobody knew what aerodynamic problems had to be surmounted, nor could they even guess what kind of stresses the airframe would need to resist. Obviously an entirely new kind of control system would be necessary. And how on earth could a jet engine be made powerful enough? The whole idea was revolutionary. Even more incredibly, the contract was not handed out according to the time-honoured method of Buggins' turn to one of the big established companies like Avro or de Havilland or Hawker. On the contrary, it was entrusted to Miles Aircraft Ltd: a minor manufacturer of excellent light aircraft and trainers.

The two Miles brothers, inspired by this extraordinary commission, came up with an equally revolutionary design: a bullet-shaped fuselage and straight, sharp-edged wings with swept-back tips. No less revolutionary were details such as a swept-back, all-moving tailplane. Hitherto, only the elevators – large tabs on the horizontal tail surfaces' trailing edges – moved up and down in order to make an aircraft go up or down. Now, however, the entire horizontal tailplane itself pivoted. This was eventually to become a standard design feature on high-performance aircraft but in early 1944 it was unheard of. The cockpit was a self-contained capsule at the tip of the 'bullet', its windscreen curved to fit the conical shape and offering no resistance to the airflow. Frank Whittle sketched plans for an engine with vastly more power, the Special Whittle W.2/700, by inventing the principle of the 'reheat tailpipe' or afterburner. In this, raw fuel is injected into the gases leaving the turbine in a jet engine in order to reheat them, providing a considerable boost in their velocity and hence more thrust. (This novel idea was also quickly to become standard in military aircraft everywhere for when extra thrust was needed, such as on take-off or in combat, its main drawback being the much increased fuel consumption.)

The Miles brothers' aircraft was known as the M.52. In great secrecy its development proceeded rapidly – amazingly so, given the exigencies of wartime, the lack of a supersonic wind tunnel anywhere in the country and the extreme complexity of the requirement. By early 1946 a full-scale mock-up of the M.52 stood in Miles's well-guarded sheds. The design was virtually complete, all the necessary materials to build three prototypes had been assembled and a first flight was planned for the summer. The M.52 was set to make aviation history.

And then in February 1946, without the least warning, a curt note from Sir Ben Lockspeiser, MAP's director general of scientific research, informed Miles that the whole thing was off. The M.52 was to be cancelled forthwith. The Miles team were stunned and disbelieving.

After all the official encouragement and support and without any apology or explanation their work was suddenly at an end. At the time, few would have guessed that the heart-stopping brutality with which the blow had fallen was to be repeated many times in the next thirty years of British aviation. The same went for the almost superstitious meanness with which the Ministry promptly ordered the immediate destruction of the mock-up as well as of all the M.52's jigs and tools. The idea was presumably to drive a stake through the project's heart to ensure it could never rise from its coffin. Frank Whittle, now an air commodore, resigned in utter disgust from Power Jets (R&D) Ltd, the turbojet development company he had founded but which had effectively been nationalised as part of the war effort. After more than two years' brilliant work Miles's hand-picked team was broken up in bitterness and disillusionment. Thanks to the Official Secrets Act the public knew nothing of the M.52's existence, let alone its cancellation, until seven months later in September.

The strange thing is that to this day it is still not completely clear why the M.52 was cancelled just when it was so close to flying. True, Britain in 1946 was chronically strapped for cash. Miles had already spent the equivalent of £2.5 million at present-day values on the project and had estimated it could require a further £250,000 (£8 million today). On 20 February Lockspeiser wrote a memo in his own hand:

We must cut our losses and cancel the contract on this aircraft . . . I have discussed the matter with the firm. There will be no tears anywhere except perhaps at PJs [Power Jets] –

but we are not paying £250,000 to test an engine. I believe the conception behind the decision to build this aircraft was to get supersonic information. We now know that was putting the cart before the horse. No more supersonic aircraft 'till our rocket propelled models and wind tunnels have given us enough information to proceed on a reliable basis.[8]

Still, the main argument against proceeding with the M.52 seems to have been about something other than money. RAE boffins at Farnborough had been studying what the Germans had been doing with swept-back wings. The first German operational jet fighter that had just preceded the Meteor, the Messerschmitt Me.262, had wings with a slight sweepback, and captured German data appeared to show that sweepback might be essential for supersonic flight. It has been suggested that the boffins didn't believe the M.52's thin straight wings would do the job.[9] If so they were wrong, as the Americans' Mach 2-capable Bell X-1 was to show only the following year: a rocket-powered experimental aircraft with thin straight wings that needed to be launched at altitude from a bomber.

Even at the time, another reason for the M.52's demise was thought to have been the influential opinion of Barnes Wallis. He was the inventor of the 'bouncing bomb' that had been used in the famous Dambusters raids in 1943 (accounted a great strategic success for the purposes of propaganda) which had earned the scientist much public acclaim. Privately, Wallis was still deeply shocked by the casualties sustained by RAF Bomber Command that night. The fifty-three aircrew lost in the raids on three

German dams (a forty per cent fatality rate) seemed to him too high a price for the result. He now expressed the opinion that supersonic flight was unlikely to be possible anyway, that test pilots' lives should not be risked in supersonic prototypes, and that all testing should be done using model aircraft. This was no doubt a humanitarian line to take but it seems badly at odds with prevailing attitudes, both in the RAF and MAP. Concerns about pilots' safety had seldom seemed to play much part in testing new aircraft designs, as a good few test pilots would have confirmed with bitter humour. Furthermore, the pilots themselves mostly adopted the sort of bravado that suggested the aeroplane had not yet been built that they couldn't master. 'I'll get the damned thing into the air if it kills me,' a test pilot might say, pausing only to pee on a wheel to bring himself luck before climbing into the cockpit. More recently it has been suggested that a better motive for the M.52's scrapping was the calculation by RAE scientists that the thrust from its intended engine, the W.2/700, would simply not have been enough even with reheat.[10] This would surely have added weight to Lockspeiser's verdict.

Whatever the reason or combination of reasons for the M.52's cancellation, it caused an uproar at the time that still echoes today as a lament for a grievously lost opportunity. Once they knew of it the press attacked Lockspeiser's decision, dubbing it 'Ben's Blunder'. The MAP had by now been subsumed into the Ministry of Supply, and it fell to the MoS to classify the M.52 officially as 'dead research', thus giving the impression that from the knowledge gained during the project's development the UK had acquired an easy mastery of the problems of supersonic flight. This was deeply disingenuous. At that time Britain knew nothing whatever of any flight faster than the Meteor that Group Captain 'Willie' Wilson had flown to his briefly held world record of 606 mph in November 1945. Well over sixty years later the whole M.52 episode looks more than anything like a brilliant and radical idea betrayed at the last moment by an equally radical loss of nerve: a sad foreshadowing of things to come. Even if it had not met its 1,000-mph design specification, the knowledge and experience gained of transonic aircraft design would surely have justified this amazingly ambitious project. As a commentator caustically remarked: 'At one stroke Britain had opted out of the supersonic manned aircraft race.'[11] 'Britain could have – and should have – been the first nation to break the sound barrier in manned flight,' another said with equal regret.[12]

Two further ironies wound up the M.52 affair. The first was that in a vicious final gesture the MoS ordered Miles to turn over all their plans, blueprints and hard-won wind-tunnel data to the Americans, who promptly capitalised on them. The rocket-powered Bell X-1 project already taking shape in the US bore considerable resemblance to the M.52, especially in its bullet-shaped fuselage and flush cockpit. This was not a matter of plagiarism so much as a convergence in aerodynamic thinking of the period. Nevertheless, the X-1 was greatly aided by Miles's data on the aerodynamics of thin straight wings at supersonic speeds, to say nothing of the invention of the all-moving tailplane. (Chuck Yeager, the USAF pilot who first went supersonic in the Bell X-1 in 1947, was to describe the all-moving tailplane as the single most significant contribution to supersonic flight.) The second irony was that Barnes Wallis was allowed to spend half a million pounds – five times more than Miles had spent – on making and flying rocket-driven scale models of the M.52, the eighth of which finally worked in October 1948. This was last seen screeching off towards the Scilly Isles, having achieved Mach 1.38 in level flight. It can have been small comfort to the Miles team to have their design officially vindicated two years and eight months after it had been so brutally cancelled.

Meanwhile, in that same spring of 1946 a strange, futuristic little aircraft began to moan around the Suffolk sky above Woodbridge. Its neat, swept-back wings flashed their metallic surfaces in the sun and onlookers noticed it had no horizontal tail at all, merely a raked-back vertical fin. They had seen nothing like it in the air before. This was de Havilland's experimental DH.108, which by the time it hit the newspaper headlines was graced with the name Swallow, although less respectful locals knew it as 'the bat jet' or 'the whistling boomerang'. They might have been even more dismissive had they realised that this first, low-speed, prototype had been somewhat cobbled together simply by tacking a pair of moderately swept-back wings on to a stubby Vampire fuselage and sticking a tail fin above the jet tailpipe. It was nevertheless classified 'Top Secret' and, to keep it away from prying eyes, was moved from Hatfield to Woodbridge, where in any case the runway was longer.

This aircraft represented de Havilland's reaction to the remarkable German Komet, the Messerschmitt Me.163: a highly advanced wartime design by the talented Dr Lippisch to which the British research aircraft bore a generic resemblance. The Komet was little more than a small rocket-powered flying wing with a tail fin but it turned out to be the Second World War's fastest combat aircraft, although the biggest toll it took was of its own pilots. Examples of the Komet captured in Germany at the end of the war, together with the aerodynamic research that had been conducted on it in Volkenröde's high-speed wind tunnels, were of particular interest to de Havilland's designers. This was because they were already working on their big DH.106 project: a tailless, very fast long-range airliner to be powered by four of their own Ghost jet engines. 'Flying wing' tailless designs were much in vogue at the time. The theory was that an aircraft without the usual horizontal stabilisers at its tail would be relieved of the weight and drag such tailplanes caused, while much larger control surfaces on its wings could make up for the absence of conventional elevators at the back. This 'flying wing' design had actually been pioneered as early as the First World War by an Irishman, J. W. Dunne (more widely known as the author of *An Experiment With Time*). From then on, a good few tailless 'flying wings' had been produced, of which the Me.163 was the fastest. The most impressive in terms of size was Northrop's immense YB-49 bomber. Aerodynamic theory said that the problem of fore-and-aft stability caused by the absence of a horizontal tail could be overcome by crafty design of the wings, although the paradox remained that if you want any aircraft to be highly manoeuvrable it needs to be less – rather than more – stable. (The inherent instability of a modern 'flying wing' design like that of Northrop's B-2 Spirit stealth bomber is such that its control relies on constant computer input to keep it in the air.)

It was precisely such parameters of stability and agility that the little DH.108 flitting around above Suffolk was testing in 1946. De Havilland soon decided that the tailless design was not suitable for scaling up into an airliner, so they changed their DH.106 drawings drastically into an early version of what was to become the Comet. Meanwhile, it became clear that the Swallow did have great potential for speed even though scientists at Farnborough had warned Geoffrey de Havilland Jr, its first pilot, that wind-tunnel tests on a model suggested it might be unstable fore-and-aft, especially at speed, when it could have a tendency to 'porpoise' through the air. However, this first of three eventual prototypes was a low-speed model and when de Havilland first flew it he was surprised to find it handled rather well. A second prototype was built with the same sweepback but designed for much higher speeds, and in September that year he displayed it at the SBAC show at Radlett where it caused a considerable stir. Back in 1946 most people had never seen a fully swept-winged aircraft, nor one where the jet engine was almost completely concealed. They were only just becoming accustomed to the Meteor with its square, straight wings and two large turbojets mounted conspicuously halfway along them. But the Swallow's single Goblin engine was hidden in the fuselage, its intakes streamlined into the wing roots and its tailpipe beneath the fin. To many present it seemed almost magical that this neat, graceful little aircraft streaking about the sky had no visible means of propulsion. It really did look exotically like the shape of things to come, and Geoffrey de Havilland Jr flew it fast enough to thrill the spectators, with the wing tips drawing intermittent thin lines of condensation through the reverberating air. Nobody seeing it at Radlett could have doubted that Britain was in the forefront of advanced aerodynamic research, and was surely destined to stay there.

By now, aircraft manufacturers and public alike in both the UK and the US had become thoroughly gripped by a sort of 'sound barrier fever'. As David Lean's eponymous film was to show, even influential people within the industry believed it would be physically impossible to fly faster than the speed of sound. This was because it was thought the steep increase in drag caused by compressibility would need an infinite amount of thrust to overcome, and also that no airframe could survive the mounting punishment. Likewise, no pilot would be physically strong enough to resist the buffeting and maintain control at such speeds. In a 1946 editorial *Flight* referred sceptically to 'the Elysian fields beyond the sonic barrier', presumably meaning that only the dead were expected to go there.[13] It is a measure of how powerful and enduring were the myths surrounding supersonic flight that when *The Sound Barrier* was released in 1952 almost five years had gone by in which many pilots on both sides of the Atlantic had exceeded the speed of sound by a considerable margin and lived to tell the tale. Despite this, one fiction the film helped promote – that it was necessary for a pilot to reverse the controls as he went supersonic – was still widely believed. This counter-intuitive idea, which implied that a pilot wishing to pull his aircraft out of a Mach 1 dive needed to push further *forward* on the control stick rather than pulling it back, achieved wide popular currency and I remember it well from school where we all believed it with an almost religious intensity. It really signified the sound barrier's conceptual status as a kind of brick wall in the sky beyond which the normal laws of physics no longer applied: a mystical region where all bets were off – perhaps analogous to modern speculation about what would happen if one fell into a black hole. In actual fact there *was* a highly limited truth behind the idea of reversing the controls, although it had nothing directly to do with compressibility but simply with an aircraft's wings being insufficiently stiff, especially if they were swept. It was discovered that in certain aircraft at a high enough speed the airflow over the ailerons (the movable slats on the trailing edges of the wings that are used make an aircraft bank one way or the other) could sometimes produce enough torsion to twist the entire wing in such a way that the aircraft would turn in the direction opposite to the one the pilot intended, requiring him to reverse the direction of his lateral pressure on the control stick to counteract it.

Such hazards certainly did nothing to diminish the aura of fearsomeness that attached to the sound barrier's public image, and an undeniable romance began to surround the whole notion of the Jet Age. Just as they had in the recent war, brave men still climbed daily into the cockpits of supremely dangerous flying machines; but these new experimental aircraft were unarmed, and the battles they now fought were against raw physics and the limitations of airframes and aerodynamic knowledge. The main difference was that the pilots were required to be less gladiators than technicians, increasingly expected to renounce John Justin-style high jinks in favour of sober step-by-step manoeuvres while carefully noting down instrument readings and other observations in pencil scribbles on a thigh pad or else talking into a wire recorder. The test pilots and their aircraft were regularly covered in press stories as well as in cinema newsreels such as Pathé Pictorial, and a slightly hysterical aura of excitement followed the activities of these heroes of the age as they set about tackling aviation's great challenge.

While the media talked about 'the sound barrier', the more knowledgeable had stopped thinking of the new jet aircraft as flying at so many knots or miles per hour, but rather in terms of Mach numbers. Ernst Mach was an Austrian physicist who in the nineteenth century had discovered how air became compressed in front of the tips of bullets and shells travelling at the speed of sound, and had photographed the shock waves that formed. In time the speed of sound in air became known as Mach 1 and after 1945 jet aircraft speeds were usually expressed as decimals of this. There was a hindrance to popular understanding, however, since the speed of sound in air is not a constant: it depends on the ambient temperature and decreases as this falls. At sea level on an average British day Mach 1 is equivalent to 762 mph. At 20,000 ft where the air is much colder, Mach 1 is only about 706 mph. Because of this and the decreasing density of air at altitude, any given speed has a bigger Mach value the higher an aircraft flies.

(In passing we may wonder why modern civil airliners like Boeing 747s and Airbuses don't go supersonic at 35,000 ft, where the speed of sound is down around 660 mph. The answer is that drag increases enormously the nearer they approach Mach 1, while the efficiency of their particular kind of jet engines – turbofans – decreases with altitude. These aircraft are designed to be limited to speeds of around Mach 0.82 to 0.85 in cruising flight at altitude, where their engines' thrust equals the aircraft's drag and it would be too costly in terms of power to go faster. Apart from that, any faster and the gremlins of compressibility await with buffeting, loss of control and the promise of structural failure.)

It should now be clear why it was usually at high altitude that the Second World War's faster propeller-driven fighters encountered critical problems of compressibility. At the limits of their performance in the thin upper air they could actually be approaching Mach 1, no matter what their air-speed indicators read. Suddenly, measuring a jet fighter's speed in knots or mph became largely irrelevant unless it was landing or taking off. For the rest of the time the test pilot's Machmeter became one of his most vital instruments. It alone could tell him how close he was to experiencing the potentially fatal effects of compressibility.

The phenomenon was all the trickier to deal with because it affected different aircraft at different speeds. It soon became more broadly named 'wave drag' as being more descriptive of what actually happened. This drag could form anywhere on an aircraft where the airflow speeded up, typically in order to get around a curved surface. As a result, the airflow could become *locally* supersonic even when the aircraft itself might be flying at well below Mach 1. The lowest speed at which this happened was known as that particular aircraft's Critical Mach Number. Since airflow obviously has to speed up more to get around a thick wing rather than a thin one, thick-winged aircraft tended to have lower Critical Mach Numbers. The fast propeller-driven American Lockheed P-38 Lightning had a fairly thick wing, which led to wave-drag problems at a relatively low Mach number, which in turn explained several accidents during the war when without warning a Lightning flying at speed would simply dive out of control into the ground. (The thinner-winged Spitfire had a higher Critical Mach Number of around M 0.89.) The onset of compressibility or wave drag could instantly reduce the lift of an entire wing. Because an aircraft's fore-and-aft equilibrium is a balance between the lift generated by the wings and that of the tailplanes, it can be thrown without warning into a violent pitching movement when this equilibrium is disturbed. This is exactly what happened to Geoffrey de Havilland Jr one evening in 1946 over the Thames estuary in his tailless DH.108 Swallow.

Ambitious private companies like de Havilland were extremely keen to break the sound barrier. In the stiff competition to win government contracts, having an aircraft with supersonic capability would be a priceless recommendation as well as doing wonders for potential exports. By now Geoffrey de Havilland Jr had already exceeded 'Teddy' Donaldson's 616 mph world record by a decent margin in testing the Swallow and he and his father had agreed that he should now make an attempt on a new official world air-speed record. On 27 September 1946 he took off from Hatfield for a final test before attempting to break the record the next day. His intention on this Friday afternoon in early autumn was to dive the aircraft to Mach 0.87 to check its controllability. The hope was that once a new speed record was behind him, the DH.108 might be further modified to go supersonic in a dive.

Some still believe that maybe his Swallow did just break the sound barrier before the strains on the airframe overwhelmed it. Certainly a dull explosion was reported as people saw pieces of wreckage falling from the sky over Egypt Bay on the Hoo Peninsula, opposite Canvey Island. Geoffrey de Havilland Jr's body came ashore many days later at Whitstable, twenty-five miles away. His neck was broken. The post-mortem established that this had almost certainly happened when the aircraft disintegrated, which could explain why his parachute was undeployed. Everything pointed towards sudden catastrophic airframe failure. His particular Swallow was simply not strongly enough engineered or aerodynamically suitable to achieve even transonic flight. It is virtually certain that this particular aircraft could never have reached Mach 1 in controlled flight, even though a much uprated version was to become the first British aircraft to do so in John Derry's hands two years later. Meanwhile, such wreckage as could be found in the sea merely confirmed structural failure under extreme stress, but the vital evidence of what had led up to this moment was still missing. Mysteriously, after some weeks two lady psychics contacted the company with a message that persuaded investigators to keep searching for the Swallow's primitive flight recorder, which was nothing more than a stylus that traced g forces and indicated air speed across a sheet of smoked glass. Against all odds this was eventually found and dredged up, and although much damaged by its immersion it was possible to deduce that the Swallow had suddenly begun to pitch up and down at a frequency of three cycles per second, a level of violent oscillation that would anyway have rendered the pilot quickly unconscious even if the aircraft had not broken up half a second later.[14] This was the uncontrollable porpoising that the RAE scientists had predicted.

Following Geoffrey Jr's death de Havilland brought in the man who was debatably Britain's most experienced test pilot, Eric 'Winkle' Brown, to evaluate the third prototype of the DH.108. Brown had flown a wider variety of aircraft of all nationalities than anybody else and had already bravely tried out one of the captured German Komets, whose fulminating rocket fuels would explode spontaneously when mixed in even minute quantities and on one hideous occasion had actually dissolved a pilot alive before rescuers could free him from his crashed aircraft. Brown duly flew the Swallow and pronounced it 'a killer. Nasty stall. Vicious undamped longitudinal oscillation at speed in bumps.'[15] At thirty-six, Geoffrey Jr was the second of Geoffrey de Havilland's sons to die testing one of the company's aircraft, his brother John having been killed three years earlier in a mid-air collision while flying a Mosquito. The idea of an autocratic company founder sacrificing his sons on the altar of technological progress surely inspired the character of John Ridgefield in *The Sound Barrier*, played with just the right mixture of ruthless ambitiousness and stoic grief by Ralph Richardson. Certainly Geoffrey Jr's death had been headline news and the subject of editorials whose predictable tone was that although it was a tragedy for the de Havilland family it was the price willingly paid by brave and brilliant men determined to maintain their country's lead in this new and dangerous science. There is a deliberate continuity lapse in David Lean's film, which has Nigel Patrick and Ann Todd returning from a brief trip to Cairo in a Comet (which only entered airline service in 1952) and buying an *Evening Standard* as they drive out of London Airport. Looking over Nigel Patrick's shoulder we study a close-up of the front page, clearly dated Saturday, 28 September 1946, with the headline 'JET PLANE EXPLODES. Geoffrey de Havilland Killed. Faster Than Sound?' above a mournful shot of wreckage strewn over mud flats. This permissible piece of artistic licence would have resonated strongly with audiences even six years later. What has genuine period authenticity is the way Nigel Patrick, who is himself playing a test pilot, shields his wife from the news as he silently reads the story while she drives the car and reminisces innocently: 'I remember you telling me once, one of the reasons you loved flying so much was that you found a sort of peace up there that you couldn't find anywhere on earth . . .'

Here, in any case, was the peculiar situation in 1946 where private aero companies were risking the lives of their sons and test pilots in a new generation of experimental jet aircraft while the government (which ultimately still controlled and paid them) largely withheld all moral support and encouragement, still convinced that the industry could relax until 1957, at least where high-performance military aircraft were concerned. It was a recognisably British phenomenon: the government forever tentative, ambivalent, irresolute, hedging its bets; probably wanting national glory but afraid of the risk of reaching for it in case it cost too much or the effort failed. This also marked the beginning of a struggle by Britain's private aircraft companies to free themselves from a command economy in favour of a return to competing freely for sales in overseas markets, especially to foreign air forces, while at the same time being highly reluctant to sacrifice any of the former dispensation's guaranteed perks and protection.

What was never in doubt was the matter-of-fact bravery of the pilots who conducted this kind of experimental test flying. Even though technically employed by private companies, they seem to have been almost as biddable as if they were still in uniform. Such uniforms as they were issued offered scant protection. There is a 1948 photograph of John Cunningham, Geoffrey de Havilland Jr's successor as the company's chief test pilot, climbing into an experimental high-altitude Vampire. It shows him wearing street clothes beneath white flying overalls: shirt and tie, short socks and ordinary black lace-up shoes. Under one arm are tucked a thin leather flying helmet and a rubber oxygen mask. He is beaming sunnily, for all the world a boy being taken up for his first flight. His cotton overalls would have offered him no protection against the numbing cold of high altitude, still less against *g* forces that might lead to black-out. His helmet's function was not protective but to hold the earphones that gave him radio communication with the ground and to provide attachments for the oxygen mask. And even that oxygen equipment was notoriously unreliable, it being not unknown for gauges to be inaccurate and for valves to freeze solid at altitude and the supply of life-saving gas to fail. In addition, the cockpits of such aircraft in those days were seldom properly pressurised, nor did they have ejector seats. Death could still come in a dozen ways even if the aircraft behaved perfectly and there was no mechanical failure.

Cunningham was a wartime ace whose night-fighting prowess had earned him the popular press nickname of 'Cat's Eyes' (which he detested). Despite his heroism, celebrity and skill de Havilland paid him an annual salary of £1,500, half of which went in tax. This was on the mean side even for the private sector and took no account of the high-risk nature of the job. On the contrary, it reflected a general assumption that men like him were so dedicated and in love with their job that they would almost have done it for nothing. Nor was de Havilland strapped for cash. Both the Swedish and Swiss air forces had ordered batches of their Vampires and Cunningham found himself delivering one of these to Switzerland on 27 September 1946. By sheer chance he took off from Hatfield at the same moment that Geoffrey de Havilland Jr left in the DH.108 on the test flight from which he never returned, and it was not until he was in Geneva that night that Cunningham heard the news. His reaction was typically understated and to the point. He described himself afterwards as 'Sad, but I had got inured to those situations over six years, having lost so many of my comrades. One just had to learn to carry on.'[16] The next year, seeing the firm's need for a replacement test pilot, he managed to talk a promising young man named John Derry into leaving Supermarine, where he was making a name for himself as a brilliant tester. Once installed at Hatfield, Derry became the Swallow's principal development pilot. The newest mark of Eric 'Winkle' Brown's 'killer' aircraft had meanwhile been structurally strengthened and given a new nose. It had even been fitted with an early Martin-Baker ejection seat, although nobody was willing to guess a pilot's chances of surviving a high-speed departure.

Derry was quite at ease with his role as one of the new breed of 'scientific' test pilots and embarked on a programme that, although called 'high speed', began cautiously at comparatively low ones. The idea was to go faster in increments until a Mach number was reached that would define the moment when the Swallow became uncontrollable. Following Geoffrey Jr's death, de Havilland's own calculations, as well as confirmatory wind tunnel testing by RAE Farnborough, had shown that tailless aircraft could become dangerously unstable longitudinally as they neared Mach 1, and it was Derry's job to ascertain exactly at what point this would happen. This deliberate 'pushing of the envelope' might have struck him as a bit perverse, albeit necessary, since at lower speeds the DH.108 handled and performed well and was a pleasure to fly. It only became a serious menace when pushed to its critical Mach speed, at which point a mere increase of M .005 (which a gust of wind might easily account for) could instantly generate negative g forces powerful enough to tear the wings off. That sort of speed difference was far too small a margin for any pilot to register, and it was a good example of the sort of razor's edge on which test pilots now had to balance their lives in their daily work at 40,000 ft.

In April 1947 Derry, flying in pinstriped trousers and a windcheater, took the Swallow to a new world record for the 100-km closed circuit. He went on methodically building up an aerodynamic picture of both the DH.108 and the Vampire over a whole range of speeds. It was a sensible way of tackling the sound barrier, but in terms of competitiveness it proved too dogged as long as Britain still lacked a genuine supersonic aircraft. In December the shock announcement came from the United States that Chuck Yeager had achieved Mach 1.05 in the rocket-powered Bell X-1, which couldn't take off under its own power but needed to be dropped at altitude from a Boeing B-29 'mother ship'. He had actually done it back in October but the USAF delayed the announcement for political as much as for security reasons. The political reasons included the calculation that in order to justify the enormous expense of the X-1 research programme it was essential that a military pilot should be the first to exceed Mach 1. However, many now believe that the first person to break the sound barrier was, in fact, a US civilian named George Welch flying a North American F-86 Sabre a fortnight before Yeager and audibly repeating the feat only minutes before Yeager's famous flight. Because Welch's Sabre held no certified Mach recording instrumentation the claim was, and remains, unconfirmable. If true, though, it was all the more impressive that the first aircraft through the barrier was a conventional jet (i.e. not an experimental rocket aircraft), and one that could take off and land under its own power – just as it would have been had the British managed to get there first.

Seventy years later it is hard to think these things still matter overmuch, except that the closing months of 1947 mark a definitive moment when British aviation was revealed to have lost the high-speed initiative. It wasn't until eleven months after Welch and Yeager that John Derry became the first pilot to break the sound barrier in a British aircraft when he finally took the much-modified DH.108 supersonic in a dive in September 1948 and barely lived to tell the tale. He began his dive from 45,000 ft, experiencing the familiar nose-heaviness, pitching and instability, all of which he felt he could cure by going to maximum power and steepening the dive. At 38,000 ft the aircraft went beyond the vertical, pulling a force of minus 3 g that tugged at his eyeballs and blurred his vision. The control column was rigid. The Swallow was now out of control at nearly the speed of sound, heading downwards with less than a minute before it bored into the ground somewhere in the Windsor area. As the needle on his Machmeter reached its stop at M 1.04 Derry closed the throttle, but without effect. Even by using both hands and exerting the maximum 280 lb pull of which he was later found to be physically capable he couldn't budge the control column by a millimetre – it might have been set in concrete. Windsor was now filling his windscreen and getting bigger by the second. In a kind of icy desperation he thought to deploy the trim tabs to their maximum upward setting. By some miracle they didn't tear away and at last the Swallow began to pull out of its murderous dive, some of its speed bleeding off in the denser lower atmosphere. As the Machmeter needle sank back the control column eased and he regained control of the aircraft, cruising it thankfully back to Hatfield. The Machmeter was checked and his supersonic speed confirmed; but John Derry's record-breaking flight had a distinct by-the-skin-of-its-teeth feel to it.

Even so, he wasn't the first British pilot to exceed Mach 1. In May 1948, on a visit to Muroc Army Air Field in California's Mojave Desert (the year before it was renamed Edwards Air Force Base), Roland Beamont had broken the sound barrier in the second prototype Sabre. He was only the fourth person to go supersonic. Bafflingly, the news was withheld from the British public, presumably from what the First World War pilot and aviation writer Geoffrey Dorman described two years later as 'that well-known "caginess" and silly secrecy complex from which our Ministry of Supply so severely suffers'.[17] Beamont, who had been an ace Typhoon pilot during the war, was English Electric's chief test pilot and was soon to test-fly both the Canberra bomber and the Lightning fighter from scratch. He summed up a popular – if somewhat face-saving – view when half a century later he wrote:

In this country we had lost the race due solely to gross government incompetence and interference in 1947 [sic] with the arbitrary cancellation of the Miles M.52 supersonic research aircraft, and the British aviation industry had had to continue with hazardous dive testing by de Havilland, Hawker and

Supermarine test pilots on developments in the DH.108, Hunter and Swift programmes into the early fifties. But none of these projects was ever going to result in a true supersonic fighter.[18]

As it turned out, all three prototypes of the Swallow were destined to crash. On 15 February 1950 Stuart Muller-Rowland was flying the high-speed Swallow in which John Derry had broken the sound barrier. Witnesses in Brickhill, Buckinghamshire, reported they had heard 'an explosion' but, as in the case of Geoffrey Jr's crash, it was considered unclear whether this was a sonic boom or the sound of the aircraft disintegrating at speed. The cockpit fell near Bow Brickhill church, to the south-east of what today is Milton Keynes. Muller-Rowland's body was found some way off. Less than three months later Squadron Leader Eric 'Jumbo' Genders took off in the original low-speed prototype to test its stalling characteristics with various deployments of flaps and lowered undercarriage, a particularly hazardous procedure. Somewhere near Hartley Wintney the aircraft was seen to pull out of a dive and immediately begin a spin from which the pilot evidently felt he could not recover since he baled out. His parachute never opened and he hit the ground not far from his aircraft. It was discovered that although the Swallow had been provided with anti-spin parachutes in wing-tip pods, they hadn't worked either.

In all, de Havilland's Swallow programme had been both tragic and highly successful. In its understated, shoestring way it was even heroic. It had not only provided the aircraft with which Britain had become the second nation to go supersonic, but the three prototypes' total of

480 flights supplied priceless aerodynamic knowledge that was incorporated into the Comet airliner as well as into the DH.110, whose wings were essentially larger versions of the Swallow's. This latter aircraft, one of whose prototypes duly turned out to be John Derry's nemesis, lost the RAF contract to its rival the Gloster Javelin, possibly as a result of the Farnborough disaster. However, de Havilland persevered and developed the aircraft to the point where the Royal Navy wanted it. The DH.110 duly went into Fleet Air Arm service as the Sea Vixen and proved very useful as a carrier-borne attack fighter around the world in the sixties, flying off carriers such as HMS *Eagle* and HMS *Centaur*. The little Swallows had served their purpose – but at a cost.

4

A Risky Business

Within four months of joining Gloster in the autumn of 1946 Bill Waterton had taken over from Eric Greenwood as their chief test pilot. Like John Cunningham at de Havilland he was paid an annual £1,500, less the punitive income tax of the day (in today's terms about £20,000 net: hardly a princely sum for a highly experienced man in his early thirties in the most dangerous profession of all). For this he was expected to take complete responsibility for scheduling and overseeing all the company's test and development flying as well as test flying and delivering production aircraft himself. In practical terms this meant that his daily workload was almost exclusively concerned with the Meteor, the company's current success, which they were building by the hundred to equip the RAF as it retired its outmoded Spitfires and Hurricanes. In addition Waterton was required to act as a general salesman for Gloster aircraft, being expected to demonstrate them both on the ground and in aerobatic displays. He was also responsible for running two airfields in the company's somewhat lunatic logistical set-up.

Thanks to wartime expansion, Gloster were employing some 4,000 people at three sites. Company headquarters and main production continued at the small airfield at Brockworth, south-east of Gloucester (the city which gave the company its name); but with the coming of jet aircraft needing longer runways the new Meteors were taken by road to Moreton Valence for testing before being delivered to the RAF. Meanwhile the company's design section was over at Bentham, fifteen miles from Moreton Valence and three from Brockworth in the opposite direction. Prototypes were manufactured and assembled by hand at Bentham, then dismantled and trucked over to Moreton Valence where they were reassembled for their first flights. In not dissimilar fashion, a few years later Britain's first V-bomber, the Vickers Valiant, would have to be taken piecemeal by road from the factory near Weybridge to Wisley, where its wings and tail were attached so that it could be flown off a grass strip. In the sheer irrationality of such disorganisation it is hard not to see some of the destructive forces destined to play their part in the ruination of Britain's aircraft industry. Decentralisation had made good sense during the war: it minimised the effects of bombing. But for long afterwards companies went on putting up with the expediencies of wartime that peace had made both pointless and ridiculous. Sensible rationalisation seemed to be beyond anybody's capability, whether at the level of government or private enterprise. In all sorts of ways throughout Britain, the war lingered on in the form of dull obedience to outdated routine.

Soon after his appointment as Gloster's chief test pilot Bill Waterton was to set another new record in 'Teddy' Donaldson's old Meteor, EE549. It and a second Meteor had been on static display in Paris, the French having decided to resurrect their international aero exhibition which had been famous before the war. The two British

aircraft quickly became a star attraction. In 1946 the
Meteor was the world's hottest jet fighter but few Parisians
had ever seen one flying and now the word went out
through diplomatic channels that they wanted to.
Waterton was summoned and given official *carte blanche*
to beat up the centre of Paris before flying EE549 back to
England. For any red-blooded pilot such an officially
approved public display offered an alluring opportunity
to show off his own skills as well as the Meteor's
capabilities; but the risks in performing rooftop-level
aerobatics in a capital city among tall buildings, flagpoles
and overhead wires were also daunting. Nevertheless, the
public's demand prevailed. 'The French didn't give a shit.
They were funny buggers that way,' Waterton reminisced
admiringly nearly sixty years afterwards.[1] With professional caution he first made a careful 'recce' in the de
Havilland Rapide biplane that Gloster's sales team used
for carrying spare parts. The French press meanwhile
whipped up interest. Newspaper articles headlined '*Le
Meteor, l'avion le plus vite du monde*' gave details of the
forthcoming display, guaranteeing a huge public turnout
on the day. There could be no backing down. Deeply
anxious about 'turning the Champs Élysées into a
slaughterhouse', Waterton nevertheless began his display
with a flat-out, tree top-level run the length of that famous
avenue at over 600 mph, pulling up into a steep climb that
ended in a loop. He repeated this several times in both
directions and at various speeds, sometimes the right way
up and sometimes inverted, interspersing the runs with a
series of dazzling aerobatics. 'The row must have been
frightful as the Meteor scorched and shrieked over the

city and between the buildings. For blurred split-seconds I glimpsed the halted traffic and upturned faces.' He was also lucky enough at the last moment to glimpse some steel cables attached to the Eiffel Tower as he flew around it. 'I missed one by no more than twelve feet. A bit closer and the steel rope would have sheared off a wing as easily as a hot knife slices butter.' He ended his show with a series of climbing rolls that took him out of sight in the western sky, having given his audience a spectacular display of jet-piloting skill that few, if any, of his peers could have matched.

His return to England in EE549 was planned by Gloster to be an attempt on the speed record for the inter-city journey, Paris to London. This was based on the in-flight time from the centre of Le Bourget Airport in north-east Paris to the centre of Croydon aerodrome. He achieved it in twenty minutes and eleven seconds. His average speed of 618.4 mph was two miles an hour faster than Donaldson's absolute world speed record over three kilometres a few months earlier in the same aircraft. By the time he reached Croydon Waterton was seriously worried about his dangerously low fuel and it would have made good sense to have landed there. However, he had been officially ordered to land only at Farnborough in order to clear Customs: a measure that was surely as typically British in its sheer daftness as the equally lunatic Paris beat-up had been typically French. He barely scraped into Farnborough with dry tanks. After refuelling, flying to Moreton Valence and taking a taxi to Brockworth, Gloster's weary chief test pilot reported to the general manager's office where that gentleman shook him by the hand and said 'Jolly good show, Waterton.' 'I repaired to my small bachelor flat at Cheltenham, fed myself, then went to the pictures.' It was 16 January 1947. 'It seemed as though I had made a good start to the New Year in my new job.'[2]

Bill Waterton's modest self-assessment must surely have masked a sense of considerable personal triumph. At that moment, at the age of almost thirty-two, he was arguably the world's number one in his field, being chief test pilot for the company that was producing the world's fastest and most developed jet fighter. De Havilland's single-jet, twin-boom little Vampire was not in the same class as the Meteor; Supermarine's straight-winged Attacker jet was not having much success, and Hawker were still turning out propeller-driven Sea Furys for the Royal Navy and export. The Meteor's pre-eminence was to be overtaken abroad in the following year by the North American F-86 Sabre and the Russian MiG-15; but in Britain Gloster's lead would last until 1951 when the Meteor was eclipsed by the Hawker Hunter and the Supermarine Swift. Even so, thanks to the Swift's failure and the Hunter's teething troubles, Meteors went on being the RAF's mainstay fighter well into the mid-fifties.

In 1947 the Meteor's gathering worldwide fame was a huge asset in terms of prestige and sales potential. Achievements such as the Paris–London record were used by aircraft companies as blue ribands to bedeck their sales campaigns. In fact, Gloster had already signed up another well-known flyer, Jan 'Zura' Zurakowski, to undertake some more record attempts and promote sales and he now fell under Bill Waterton's command. During the war

'Zura' had flown with the Polish squadron based at
Northolt. Even among his compatriots, who were
notorious for a style of flying that verged on the lunatic, he
had built up a reputation as an outstandingly brilliant
pilot. Like Neville Duke, the future inventor of the
'Zurabatic cartwheel' aerobatic was a graduate of the
second-ever course at Britain's prestigious Empire Test
Pilots' School and had then been employed for a while by
the MoD at A&AEE Boscombe Down.* In 1950 Gloster
told 'Zura' to attempt the London–Copenhagen–London
speed record in order to improve their chances of selling
the Meteor F.Mk 8 to the Danish Air Force. It worked: the
record was broken and the Danes bought the F.Mk 8.
Waterton always evinced great respect for his Polish
subordinate's piloting skills while being slightly more
reserved about his personal qualities. 'Never turn your
back on Zurakowski,' a colleague was once overheard to
remark.

Establishing new records and displaying aircraft at air
shows were two of the industry's main selling ploys. Letting
potential customers fly them was another. Even in early
1947 it was clear to Bill Waterton that his contractual
obligation to double as one of the company's salesmen as
well as its chief test pilot was to be no sinecure. Gloster
built a demonstration aircraft, a clipped-wing Meteor F.Mk
IV, which, once its guns had been removed, was painted a
glossy carmine and given the civil registration G-AIDC,

---

* Aeroplane & Armament Experimental Establishment. Home of ETPS, the
world's first school for test pilots (motto: *Learn to Test – Test to Learn*),
Boscombe is nowadays run for the Ministry of Defence by QinetiQ, a UK/US-
owned defence technology and security company.

thereby becoming the world's first civil jet.* It was also fitted with several extra fuel tanks in acknowledgement of the Meteor's voracious low-altitude thirst of a gallon of kerosene per mile. G-AIDC looked splendid in its dazzling red and cream scheme and, Waterton being too busy with testing, it was dispatched on a tour of Europe in the hands of some of Gloster's junior pilots. Unfortunately, it didn't get very far before it was written off in Brussels while being flown by a Belgian. What strikes one as peculiarly British about this enterprise was that the gleaming G-AIDC was accompanied on its tour by a de Havilland Rapide full of Gloster mechanics. Who else but a British company would back up a foreign sales drive showcasing a beautifully finished specimen of the world's fastest production fighter with a thirties biplane made of canvas and plywood and with a top speed of 150 mph?

I should add that the Rapide was an excellent aircraft, and specimens are still flying all over the world even today. I have a particular affection for it since it gave me my first-ever experience of flight on that twenty-minute joy-ride around London Airport in 1952. I can still remember the steep angle of the floor when I first climbed aboard (the Rapide was a classic tail-dragger) and the pale circle of the fabric escape hatch in the roof. Once aloft, I can also remember the strange sensation of watching the landscape tilt like a wall, my body fooled by centrifugal force into believing that I was still sitting level and the ground below

---

* Until June 1948 mark numbers were indicated in Roman numerals. In that month Arabic numerals were adopted (so the F.Mk IV became the F.Mk 4) and all subsequent versions of the Meteor were given Arabic numerals. (www.historyofwar.org/articles/weapons_gloster_meteor_IV.html)

was leaning. From the moment we touched down I discovered that a part of me was still airborne and it seems that it has remained so ever since.

While G-AIDC was abroad Jimmy Bridge, another of Gloster's test pilots, was killed when attempting to land his Meteor at Moreton Valence after an engine failure. He was a close friend of Waterton's and it fell to Bill to break the news to his widow. He found Mrs Bridge out shopping in Cheltenham. 'She was awfully brave about it, as the women of men who fly usually are. Looking back over fifteen years, a cruel quirk of fate always seemed to select pregnant wives to have such news broken to them.' As if this disagreeable task were not enough, Waterton was then told that as chief test pilot it was his duty to identify his friend's badly disfigured body – an unnecessarily heartless order presumably given by those who lacked the stomach for it themselves. 'It was an unpleasant task, and the sight stayed with me. But the job had to go on, and more aircraft were waiting to be flown that afternoon.'[3] Indeed, his log books reveal he was flying an inordinate number of Meteors each month, ensuring that each one produced was fit for delivery. He was also performing at air shows such as the SBAC display at Radlett. The *Tatler*'s correspondent wrote admiringly: 'Squadron Leader W. A. Waterton in the Meteor ... gave the best jet aerobatic display I have seen, swooping up in rocket loops which took him to 10,000 ft high, and then grass-cutting past the spectators at great speed.'[4] This went on until March 1948 when the process was briefly interrupted by his having to test-fly Gloster's new prototype, the straight-winged E.1/44. It was not long before he had scornfully dubbed it the Gloster 'Gormless' on account of its poor performance and shoddy finish. The project was soon dropped, not least because the design itself was obsolete. The 'Gormless' was quietly forgotten, and with it a large sum of taxpayers' money.

In view of what had happened to their beautiful sales Meteor in Brussels, the company wisely concluded that it was a bad idea to allow inexperienced potential customers to solo in their precious aircraft and decided to build a two-seat, dual-control version, the Meteor T.Mk 7, which was to become the world's first jet trainer. This first example was designed as Bill Waterton's demonstration aircraft, to accompany him abroad on his sales tours and allow him to take up clients who were pilots while he exercised his famous instructor's patience in the back seat. However, it was also the cause of the first of many serious quarrels with his employers.

From the very beginning he loathed the swing-over cockpit canopy, which weighed a prodigious 264 lb and had an unreliable locking mechanism. Worse, the rear – or instructor's – cockpit was ludicrously inadequate. When years later Wing Commander J. A. 'Robby' Robinson was to refer to British cockpits of the period as 'ergonomic slums', he was simply generalising about a feature that had so long characterised so many aircraft that it had become a point of national recognition, and one responsible for countless accidents. It seemed as though cockpit layouts were habitually designed by people who didn't fly, the instruments and controls bunged together higgledy-piggledy with very little thought as to how they might help the wretched pilot, especially in moments of crisis. The

Meteor T.Mk 7's all-important rear cockpit completely lacked elementary instruments such as fuel gauges, jet-pipe temperature gauges and even relight buttons in case of an engine flame-out. Waterton began raising these issues with the design department at Bentham, being at first fobbed off and then, as his vehemence increased, ignored altogether – 'cold-shouldered' was his phrase. It was all the more extraordinary given that his were the professional objections of the company's own highly experienced chief test pilot. His job, after all, was to fly and approve their designs. When he learned that the RAF was interested in buying the T.Mk 7 as a trainer he felt obliged to warn the Ministry of Supply's representative at Gloster, the Resident Technical Officer, that the aircraft embodied serious problems which the design team were flatly refusing to address. The resultant letter 'caused a furore at the Ministry, but it was still years before the rear cockpit equipment was supplemented'. This marked the first of many occasions when relations between Gloster and their chief test pilot became acrimonious. The company regarded his reporting to the Ministry's man as 'unprofessional, unethical, underhand – and more besides'.[5] It is a good bet that the 'more besides' included an undercurrent of the usual accusation that, despite his position in the company, he was not really 'one of us', being Canadian and ultimately, therefore, an outsider whose loyalty would always remain questionable.

Considering the company's reaction and the way in which Waterton was later to be blackballed by the entire industry, it is worth examining his position. No doubt he himself stated it undiplomatically and with characteristic bluntness, and this would inevitably have struck a grating note in a company accustomed to the chummy, relaxed atmosphere of an outfit favoured by the government and the RAF, both of whom were eager for its products. But as far as he was concerned his stance was a matter of simple professionalism. Gloster had hired him to test prototype and production aircraft to the point where he could pass them as fit to be flown by average pilots in squadron service. Although the unarmed (and hence much lighter) T.Mk 7 Meteor in fact flew beautifully, it was marred by details such as the potentially lethal defects in its cockpit design. 'I felt I had a duty to my fellow pilots in the RAF,' said Waterton: a sentiment that surely reveals how instinctively he thought of himself as still in the service that was his 'first love', and only secondarily as a company man. The doubts that some of Gloster's management evidently harboured about his loyalty were therefore probably justified, although wholly misplaced. His being Canadian by birth had nothing to do with it; his having recently left the RAF as a bemedalled squadron leader had everything. (The emotional transition from air force to private enterprise was one that several other test pilots of the era probably never made fully. It was doubly difficult when testing top-secret prototypes that might well become military aircraft. The pilots were constantly landing at RAF airfields and dealing with men in uniform – many of them former chums and comrades – who represented the potential end-users and who addressed them by the rank they had formerly held. Besides, civvy street in peacetime lacked the comradely cement that service life had provided during the war.)

The incident blew over, but the underlying situation remained unresolved. Gloster's idea that their chief test pilot's duty was more or less to rubber-stamp their designs was incompatible with his conviction that his primary duty lay in protecting the lives of those who were to fly them. It was a conflict that was to smoulder on, now and then flaring up into fierce outbreaks, for the remainder of Waterton's employment. He duly took over his dual-control T.Mk 7 Meteor which, like its single-seat predecessor, was finished in eye-catching carmine and cream and furnished with a civil registration, G-AKPK. In May 1948 he set off in it on an extensive overseas tour, selling aircraft in Turkey, Greece, Italy and France, in all of which he caused a sensation with his aerobatics, turning out to be surprisingly good at charming the brass-hats and even royalty that took such interest in him and his machine. On different occasions he flew with the Shah of Persia and Prince Bernhard of the Netherlands in the Meteor's front seat, both of whom were accomplished pilots. There are several photographs of Waterton and the prince sitting in the T.Mk 7 grinning like schoolboys.*

Apart from such exotic breaks in routine, the daily grind in Gloucestershire of testing production and experimental Meteors continued. This was the bread-and-butter part of the business that schoolboy aircraft spotters like

---

* Photographs of Prince Bernhard sitting in the cockpits of various jet fighters were to litter the memoirs of many other test pilots of the period. That was before the prince became one of the central figures in the Lockheed scandal when in 1976 it was revealed that he had demanded and received $1.1 million in bribes from the US company in return for influencing the Dutch Air Force's choice of the F-104 Starfighter as their next interceptor. The affair eventually led to the prince's disgrace and his wife Queen Juliana's abdication.

myself never really considered. We saw our heroes as perpetually breaking records while now and then dicing successfully with death in some hair-raising cockpit drama. It surely never occurred to us that they might be underpaid men who kept office hours, arriving early in the morning at cold, mist-shrouded airfields to find a backlog of twenty identical Meteors waiting, beaded with dew, each one of which had to be taken up for a couple of quick circuits to check that it had been put together correctly. Bill Waterton's respite was now and then to fly the firm's Gladiator, a sole specimen of which had been restored by the apprentices. The Gloster Gladiator was a thirties fighter biplane that had become famous in the stubborn defence of Malta in 1940. After the 'powered bricks' that to him were modern jet aircraft, Waterton found the vibrating old Gladiator took him straight back to the exhilaration and discomforts of his own apprentice-ship as an airman ten years earlier. He would perform endless aerobatics over Moreton Valence and revel in the gale that poured into the cockpit. 'This was really *flying*: a joyous thrill I had forgotten during my past four-and-a-half thousand hours in the air.' It might be showery, but he flew with the canopy open because 'clouds and rain were not enemies, but sweet manifestations of nature that made you feel clean and young and exuberantly detached'.[6] 'Exuberantly detached' is the perfect aviator's phrase and might have been coined by anyone who recognised the sublime poetry of being able to turn the boundless sky into a private playground from a tiny open cockpit.

Our house in Kent lay barely eight miles from Biggin Hill,

that famous Battle of Britain airfield. When we moved there in 1948 Spitfires and Hurricanes were still carrying out mock dog fights like whirling midges high above our garden, but they were rapidly being edged out by Meteors. We were directly beneath a regular flight path and the sight and sound of the jets coming over on approach was a daily pleasure of which I never tired. In fact, when I was eight and sent away to a boarding school in Sussex, one of the things I most missed amid the vile silence of the Sussex Downs was the friendly, honking roar of the Meteors' twin Rolls-Royce Derwents as they cruised in over our roof at probably no more than 120 knots. I particularly liked the characteristic warble sometimes produced by a freak of acoustics as they disappeared behind the tall elms that lined the garden. I could tell the different types at a glance. They were mostly F.Mk 8s, but an occasional NF. Mk 11 came over – the night-fighter version with its lengthened nose containing radar. I also became familiar with the T.Mk 7: the two-seat trainer which had the long cockpit canopy whose gross unwieldiness had aroused Bill Waterton's opposition. To my child's eye it looked as though it was glazed with panes of glass like Crittall windows.

In my memory the vast majority of air activity in my boyhood was military. Compared with today there was limited civil aviation, and until the Comet flew any jet aircraft was de facto military. In my childish naivety the aircraft I watched and hankered after existed for themselves rather than being designed for a specific task. They just *were*: fast, noisy and blood-tinglingly spectacular. One summer my father took me to an air display at Biggin Hill.

One of the sideshows that year was a Meteor F.Mk 8 parked facing the gun-testing butts. For a shilling one could line up to sit in its cockpit and fire a single round from one of its four 20 mm cannon. There was a rather tame ladder leading up to the cockpit, manned by a uniformed flight sergeant, which was a bit disappointing because I had wanted to climb up that slab metal side as I had already seen a pilot do, pushing his feet into spring-loaded steps that closed up after him. Still, it was an inexpressible thrill to sit in the well-worn military seat in a cockpit that was surprisingly smaller than I had imagined. I tried in precious seconds to take a mental photograph of everything. It didn't seem possible that the miracle of flight could be controlled from this cramped and slightly tatty den of chipped black paint with the cracked rubber hot-air tubes for demisting crudely anchored to the metal windscreen surround with what looked like duct tape. The dials and switches were set in matt-black panels, some of them backed by little patterns of diagonal yellow and black stripes. 'There's others waiting, you know.' A blue serge arm reached in and showed me which button to press on the control stick (which I also noticed had something like an upright bicycle brake lever behind its grip). So, glaring with Biggles-like intensity through the gun sight at the sand dune as though I actually had some choice in my aim, I thumbed the firing button and felt the aircraft shudder in response to the loud explosion somewhere down by my left leg. A simultaneous fountain of sand erupted several yards away, and that was that. I was helped back down the ladder and allowed to keep the empty cartridge case, still hot and aromatic with the peppery scent of fresh cordite. I spent the rest of that afternoon in a daze. *I had felt the power.* I had actually sat in a jet fighter and fired its gun. From then on I lived a good deal of my daily life inwardly at several thousand feet, peering ahead through a Meteor F.Mk 8's V-shaped Triplex windscreen, my left hand on the throttles and my right lightly on the control stick, my eyes flicking towards the Machmeter top left on the instrument panel. This view was a lot more vivid to me than were the dreary ground-level realities of classrooms and cricket pitches. The bulky brass 20 mm cartridge case became a talisman that went everywhere with me. I was in no doubt where my future lay and eagerly pored over the advertisements for a career in the RAF that appeared regularly in magazines like *Boy's Own*.

What was it like to fly those thrilling lumps of metal with their somewhat unreliable ejector seats? An RAF pilot officer, newly qualified on propeller aircraft and fresh out of Cranwell with his 'wings' up, would arrive somewhere like the Advanced Flying School at RAF Oakington in Cambridgeshire, acting laconic while secretly tingling with boyish excitement. Here at last he would be turned into one of those exalted creatures whose ranks he had long been aching to join: a jet pilot. Some of his tutors at Cranwell had been ex-Battle of Britain pilots and, of course, they too were gods in their way. But they were ageing gods. Some were in their mid-thirties, others even older. His was a new generation born to a new kind of aviation. As the ads said, there was a future for him in the RAF. (See the young ace gazing up at the sky where his destiny lies scrawled in contrails left by gleaming specks of hurtling technology!) At rare moments of sober introspection he would probably admit that these early jets he was about to fly, such as the Gloster Meteor, were still on the crude side even if they did represent the state of the art, and a pilot's life expectancy could be short . . . But those were just *statistics*, which everyone knows can be used to suggest anything. The aircraft were improving all the time and luckily lads like him were immortal. It was always the other poor sod who pranged and burnt, as he'd noticed at Cranwell when both Jimbo and Stubbsy had spun out in their Harvards. (On this boneheaded quirk of male psychology has the entire history of human warfare depended.)

The bus driver who brought the young 'Robby' Robinson and his fellow graduates to Oakington in January 1953

had ways of cutting cocky young officers down to size. He did not take us through the main gates but turned into a back entrance that led through the station scrapyard. Now he chose to go slowly. We gazed around us in silence. There were crashed Meteors as far as the eye could see. Many were piled up like corn stooks, their tails resting against each other. Others were too far gone for this and were just piled one upon the other . . . I was shown to my room by an airman batman who did not offer to carry my bags. We stopped and the airman produced a key and opened the door. I noticed that around the doorframe hung several pieces of string, held there by black sealing wax. Even I knew what these meant: the previous occupant had been killed, and recently too. I peered around the darkened room as my guide drew back the flimsy curtains.

'Sorry about the seals, sir,' he said. 'They should have taken those down.'

I looked around to see if any traces of the previous occupant remained. There was nothing. A life had been wiped away as if to make room for me.[7]

Black sealing wax. Somebody must once have dreamt that up. In whichever department of the MoS or RAF responsible for that service's day-to-day supplies, requisitions would now be regularly arriving from operational squadrons all over Britain: Sealing wax, Black, Quarters of deceased aircrew for the securing of.

As a jet-struck boy I lived daily with an image of myself as an RAF pilot walking out at that tingling moment in early light towards a line of parked Gloster Meteor F.Mk 8s, their canopies still misty with dew which the ground crews were sponging off in readiness. It was the details of the scene that counted: the paling sky, the beaded cobwebs in the grass at the edge of the hardstanding, a Bedford maintenance truck returning along the perimeter having doused and refilled the paraffin gooseneck flares that were still used as back-up in case the runway lights failed. Then climbing up into the cockpit, which was cold and full of the faint perishing-rubber smell of the canopy seating and the insulation of the hundredweights of wiring harness concealed behind metal panels and bulkheads, some of which would give off a hint of phenolic resin as the circuits warmed up. Then the strapping in, the stowing of the bang seat's safety pin and the serious, never-to-be-skimped pre-flight checklist on which your life really would depend in the next hour. Close and dog the canopy. Test the radio channels – control tower and flight leader – and get the go-ahead to start engines. Your gloved hand raised to the erk manning the starter trolley. Press engine number one starter button and listen to the vacuum-cleaner whine mount as the turbine gains speed. Wait for that moment when the starter motor is turning over at maximum and the green landing gear lights dim. Then slowly push down the cock that squirts high-pressure kerosene into the rotating engine. The igniter plugs crackle and the heart-quickening smell of paraffin fills the cockpit. Watch the jet pipe temperature gauge until it's safe to push the cock fully down and hear the left-hand Derwent roar steadily. Repeat the process with the starboard engine. Wave away the starter trolley and the chocks and listen out for permission to taxi. Throttles eased forward enough to begin rolling and gun the port engine to swing onto the taxiway. Check the windsock on its pole out of sheer habit. This morning it's hanging motionless. The damp brakes squeal as you halt for take-off clearance. Then throttles fully forward, briefly check both rev counters and jet-pipe temperature gauges, brakes off and concentrate on not swinging before climbing up over the numbers at the runway's end, four tons of aircraft miraculously borne aloft by the delicate dawn air. The rapture of the morning sky with its fluffy furniture and the bravura technology that gave you access to it: twin marvels that you never could separate. I didn't doubt that there was a mystical pact between them.

It was, however, a pact that could be abrogated at a second's notice when either the air betrayed the airman or the airman failed the air. Often, pure dumb luck decided a pilot's fate and aviators knew well that no amount of skill, prayers or talismans could make a scrap of difference if your number was up. Take the case of Douglas Wakeman, a test pilot for the Air Registration Board, who was flying a Bristol Brigand on a routine test one day. This was the RAF's last propeller attack aircraft: a twin-engined post-war successor to the Bristol Beaufighter. Wakeman was flying along when without warning the starboard propeller detached itself. Still spinning, it fell below his aircraft and then rose on the other side, colliding with the wing and shearing off the entire port Centaurus engine. Although the aircraft was now mortally damaged and completely without power, Wakeman managed to glide the Brigand down to a perfect dead-stick landing in a field. It was a brilliant piece of cool-headed flying, but at the last moment he lacked the engine power to swerve at the end of his run to avoid hitting a tree which toppled across the cockpit and killed him. The god who watches over airmen shares the same sense of humour as the deity in a Thomas Hardy novel, and they know it.

The ways in which you could lose your life in an early jet like the Meteor were legion. Apart from the already-mentioned 'phantom dive' that could bring disaster on approach if you forgot to retract the air brakes before lowering the wheels, there were the cockpit canopies that sometimes burst or blew off at speed. These might either decapitate you outright or merely flash-freeze your head and lungs. One pilot survived this at the expense of chronic lung damage that rendered him an invalid for the rest of his life. Meteors also suffered badly from cockpit icing and misting. It was enough to take them to altitude and then dive, a manoeuvre any fighter aircraft might be expected to perform. As they hit the lower, warmer air the canopies went opaque with ice outside and frost inside, leaving the pilot suddenly blind. The challenge was to see enough in order to be able to land. At the time there were practically no ground-based navigation aids and the early turbojets were notoriously thirsty. In a Meteor, for example, once you were in the air you had about forty minutes' flying time ahead of you – less at low altitude – before the fuel ran out and you had to put your four-ton 'Meatbox' down somewhere, *any*where. With an opaque canopy it could turn into a desperate struggle, flying with one hand in snatches between trying to keep a patch of window clear with the back of a glove, one eye straining for visual clues of a possible nearby airfield, a second eye on the air-speed indicator and altimeter and a third eye on the twin fuel gauges, which meant some quick refocusing since they were ridiculously sited down past your knees at the base of the control column.

Not infrequently it could all go badly wrong, as it would on a famous occasion in late 1953 when four Meteor F.Mk 8s from 56 Squadron were trying to get back to base at RAF Waterbeach in Cambridgeshire. Visibility was bad with low cloud and fog and all four aircraft were dangerously low in fuel. Flying in two pairs, they made several attempts to get into Waterbeach but the airfield still lacked a GCA system.* Utterly lost, one pair made the wise decision to use their last pints of fuel to climb up above the clouds and eject. Their Martin-Baker bang seats

---

* Ground-controlled approach. The radar-based landing system on which Arthur C. Clarke worked during the war as an RAF radar technician.

duly blew them out into the sunshine and they parachuted safely down while their aircraft thudded deep into the wet fenland below. The other pair left it too late and ran out of fuel, instantly dropping like anvils. One pilot ejected at 900 ft, which was below the safe minimum height for the Mk 1 seat. His parachute opened just in time although he badly injured his back on landing. His companion was not so lucky and tried to land his aircraft in a fog-bound field. In the ensuing crash he sustained injuries that left him crippled for life. Four aircraft had been written off in as many minutes.*

It may well seem strange to us, over half a century later, how little effort designers in those days appear to have put into making their products more user-friendly for the pilots expected to fly them. After all, depending on its type each Meteor cost around £38,000 in 1950 (almost £1 million today) and pilots were expensive to train. Probably not much could have been done in a hurry to improve engines and avionics or, come to that, instrument landing systems: no doubt these had to evolve at their own pace. But basic things like cockpit pressurisation, heating, de-icing and instrument layout were not beyond the capabilities of current technology, and pilots' memories of what they had to put up with often remain vivid to this day. Much of the flight testing of the various Meteor marks was quite hazardous enough without the additional problem of finding yourself rendered blind by ice on your

---

* A similar but even worse instance was to occur in early February 1956, when six out of eight Hawker Hunters were caught in fog over East Anglia, ran out of fuel and crashed. One pilot was killed and the cost of the aircraft alone was estimated at £500,000 – getting on for £10 million at today's values.

canopy. The recommended expedient was to accelerate to over 450 knots in the hope that the kinetic heating effect of air friction would clear a patch of vision. The penalty for this was that you used up still more of your precious fuel.

Throughout my childhood and adolescence there were almost daily press reports of air accidents. I naturally longed to see an aircraft crash but never did. I lived in constant hope. Everybody knew of the famous day in 1951 when three Meteors had crashed fatally just up the road at Biggin Hill, all within a hundred yards of one another. Infuriatingly, I was away at school in 1956 when another Biggin Hill Meteor crashed within three hundred yards of our house. It came down in some school playing fields and exploded. By the time I returned home for the holidays every last souvenir fragment of aluminium and perspex had long been looted. The incident was remembered in our family because the pilot had been taken to Queen Mary's Hospital, Sidcup, where by chance my mother was duty anaesthetist for that day, so he became her patient. The story was that the Meteor was seen to be low and in trouble and the pilot had ejected. His partially opened parachute had snagged in the trees at the edge of the field. When found, he was in shock with a badly broken leg. He died the next day in Queen Mary's. My mother's explanation of how a broken leg had killed him was that the impact had sent quantities of bone marrow in suspension into his bloodstream and this emulsion had presumably formed some sort of fatal embolism. He might also have had internal injuries; I can't remember.

When I began writing this book I suddenly wanted to find out more about this incident which, after upwards of a half-century, had acquired for me an almost mythical quality as of something dreamed. A bit of foraging revealed that the *Sidcup & Kentish Times* of 19 October 1956 could fill out some of the story. The pilot was 22-year-old Flying Officer Roger Coulston from 41 Squadron, Biggin Hill. There was much talk of his courage in having 'stayed at the controls of his crashing Meteor until it was clear of houses', although this sounded like one of those obligatory *nil nisi bonum* formulae for a dead airman, given that the last mile he had flown had been over open farmland. Eyewitnesses said 'the plane appeared to get into difficulties when it dived to the left and spun upside down. As the plane rolled over the canopy came off. Then the ejector seat and parachute came out.' There was little more to be learned from the local paper except that the RAF had thoughtfully flown the pilot's mother up from Devon and that she saw her son before he died. It was up to the records of the Accidents Investigation Branch in the National Archives to add further detail.* These AIB records are concise and laconic. Accidents to RAF aircraft in the fifties were so frequent there must have been teams of crash inspectors permanently whizzing about the country examining grim messes. Barely abridged, the report into Flying Officer Coulston's crash reads as follows:

---

* In late 1987 the AIB became the AAIB: the Air Accidents Investigation Branch. Since 2002 the AAIB has been part of the Department for Transport (DfT).

Accident to Meteor Mk. 8 WA 855 on 13th October 1956

1. The aircraft was on its initial approach to Biggin Hill under GCA direction . . . At about 200 feet the canopy was jettisoned and the pilot ejected . . . The pilot, who was seriously injured, died the following evening without being able to give a statement. However, the Police Constable who rendered first aid has said that the pilot told him that 'The aircraft got into trouble. I pulled the ejector seat and baled out. That is all I remember. I left it late in baling because of the houses.'

2. (i) The aircraft was built in 1950 and had flown 1,281 hours since new.

(ii) The pilot, aged 22 years, held an 'Average' flying assessment and a White Instrument Rating. His total flying experience amounted to 361 hours as first pilot, most of this time being spent in Vampires and Hunters. His experience in the Meteor amounted to only 5 hours 20 minutes. He was last checked out in a Meteor 7 in July 1956, since when he had flown only four sorties in the Meteor 8. He had recently been suspended from flying for 7 weeks pending Court Martial, and on his return to duty on the 11th October he was given a satisfactory dual check on a Vampire T.11.

(iii) No mechanical failure noticed. Martin-Baker Mk. 2E ejector seat had functioned properly but the aircraft was too low for the complete development of the parachute.

3. Conclusions: The accident was probably due to loss of control whilst flying very slowly during a GCA descent . . . The pilot's general inexperience in the type, and his lack of recent flying experience in the Meteor is noted.[8]

It seems likely that Coulston quite simply stalled the aircraft. Basically, he was too low and too slow. At that time it was RAF practice for Meteors to do approaches at 100–110 knots, which was really much too slow for safety because if an engine failed the aircraft needed to be doing at least 125 knots to have any hope of recovery. This young pilot was very average (his 'White' instrument rating category was the most basic, from which he might have hoped eventually to progress to 'Green' and even to the godly status of 'Master Green'). He was also relatively inexperienced on the type. It is likely that, since he had been suspended from flying, his court martial was for an aerial offence such as low flying. Anyway, I salute the poor lad across the decades: *mon frère, mon semblable* of my boyhood fantasies (although thanks to my own Biggles-like skill I was naturally never going to get killed). Young Roger Coulston was the typical average RAF pilot of the time, neither very expert nor experienced: exactly the sort of person that conscientious test pilots like Bill Waterton had in mind when trying to remedy a new aircraft's faults before passing it as suitable for squadron use.

Since this was written in 2009 Coulston's nephew Dan has conducted his own researches into his uncle's death. It turns out Coulston was never properly converted to Meteors by a Qualified Flying Instructor but had merely been taken up by fellow pilots in a T.7 trainer for a bit of practice. His main experience was all on single-engined jets, the Vampire and the Hunter (his recent grounding had been for low-flying a Hunter along the South coast). The point was that he had very few hours on twin-jet aircraft and should never have been cleared to fly one that day. His nephew has discovered that as Coulston was on approach to Biggin Hill the air traffic controller radioed that there was another aircraft in the circuit needing to land before him, and told him to go around. While doing that, it looks as though Coulston may have taken the opportunity to practise asymmetric flying with a simulated failure of one engine – a notoriously dangerous manoeuvre in the Meteor that had killed many pilots, including instructors. If so, it is even less surprising that he stalled and became one of the 444 Meteor fatalities between 1944 and 1986 – a period in which a total of 890 Meteors were lost.*

A quick glance at the rest of this particular AIB file reveals many similar casualties. It deals with accidents that took place between 3 May 1956 and 3 January 1957. In those mere eight months there was a total of thirty-four accidents in which forty-two aircrew were killed (roughly one fatality every six days). Pilot error and mechanical failure shared approximately equal billing in the official list of causes. The aircraft types included ten de Havilland Venoms, six de Havilland Vampires, six Hawker Hunters, four English Electric Canberras, two Gloster Meteors, and one each of the following: Gloster Javelin, Folland Gnat, Avro Vulcan, Avro Shackleton, Short Seamew and Westland Whirlwind helicopter. Three of the accidents involved test pilots. The Seamew was flown by Squadron Leader W. J. 'Wally' Runciman, Short Brothers and Harland's chief test pilot, who looped it into the ground during an air display at Belfast's Sydenham Airport on 9 June. That was the graver of two miscalculations on his part, the other being ever to have flown such an aircraft in a display. The Seamew had been designed in 1951 as a

---

* See Nick Carter, *Meteor Eject* (2000).

cheap Anti-Submarine Warfare aircraft for the Royal Navy. It was an absurd-looking machine with its tall, narrow fuselage and weirdly hunched cockpit perched on top, seemingly within inches of the propeller. Only nineteen were ever built before the project was axed, and a later description of it as 'a camel amongst race horses' was, if anything, generous.[9] What was a chief test pilot of Runciman's experience doing looping a dud aircraft that had no future and that not even Short could have been proud of? To lose one's life doing a silly thing in a bad aeroplane is a sad end to an accomplished career, but it was hardly rare in those days. The Folland Gnat, on the other hand, crashed because of structural failure and luckily their chief test pilot ejected safely. He was Squadron Leader E. A. 'Teddy' Tennant, who lived to comment publicly on an Avro Vulcan crash two months later on 1 October: the famous Heathrow disaster of 1956. More than thirty test pilots lost their lives in the aircraft they were flying in the first six years after the war alone.[10] This toll, as well as that of ordinary RAF pilots, steadily rose throughout the fifties.

In late 1949 Bill Waterton was abruptly recalled from a sales trip in Egypt to find that Sir Roy Dobson, then a director and soon to be chairman of the Hawker Siddeley Group, had decided he was to be sent to Canada. The idea was that Waterton should be seconded to Avro Canada, another company in the Group, which had just finished the prototype of their first jet fighter, the CF-100. The task of flying it was to fall to Waterton on the grounds that there was no test pilot in Canada with anything like his experience on twin-engined jets. Dobson, who as the wartime head of A. V. Roe had personally ordered and supervised the production of the Lancaster bomber, was a considerable autocrat to whom nobody dared say 'No'. Bill Waterton, foreseeing trouble in store for him in Canada, promptly said 'No' but was quickly brought to heel. What he didn't know was that behind all this lay an ambitious company strategy. Dobson, who had visited Canada during the war and had been highly impressed by the Canadians' energy in running their branch of Avro, was now planning to use Avro Canada as the means to enable the Hawker Siddeley Group to challenge US supremacy and become a major competitor in the North American market. So Waterton, a mere employee, obediently packed his bags and returned to his native land full of misgivings. He sensed that he might be putting some Canadian noses out of joint, and he was quickly proved right.

A few days after he arrived in Toronto he was given a lift in a stretch limo full of businessmen and Avro Canada investors in black coats and homburg hats. 'I'm riding along and I don't look like a Canadian because I'm wearing my hair as it has always been and I've got my old grey coat that my tailor had turned and I still had a moustache, which wasn't common in this country. And these people were talking away and one of them said, "I hear they're sending a Limey over here to fly our aircraft. We've got Billy Bishop* and people like that – who can beat *them*?"

---

* William Bishop VC was the Canadian flying ace from Owen Sound, Ontario, who scored a record seventy-two victories in the First World War. A national hero, he was made an honorary air marshal and directed RCAF in the Second World War. He died in 1956 and thus was still alive at the time of this conversation.

So this was my introduction to the company . . .'[11] It was not much of a welcome for a native Albertan, both sides of whose family had been in Canada long before his father's Ontario and his mother's Alberta had even been defined as provinces. It was surely all the more galling for someone who had served in the Canadian military for four years before going overseas to fight Canada's war in Europe.

The CF-100 Canuck that Waterton had been sent to test was an excellent design for its day. It was a two-seat long-range all-weather fighter with twin Rolls-Royce Avon engines mounted close to the fuselage. Straight-winged, it nevertheless had a good turn of speed and, unlike the Meteor, could be pushed to go supersonic in a dive. However, that sort of treatment had to wait a long time while Waterton took the prototype through its early paces. The first flight was on 19 January 1950, and almost at once he encountered problems with the undercarriage. Such things were not unusual in a new aircraft and it is a measure of how quickly he felt confident of its handling that the third flight was witnessed by Sir Roy Dobson himself, who had flown to Canada to see what the aircraft would do. Even though it was only the Canuck's third time up, Waterton risked 'a mild beat-up. Nothing elaborate: just high- and low-speed flying, with rolls and tight turns.'[12] The show went down well; but when he landed it was discovered that the fairings at the aircraft's wing roots were torn and twisted. Further investigation revealed that the whole of the centre section and the main spar were far from strong enough. This led to major redesigning which slowed things down, although, as Waterton acidly noted, nothing like as much as it would have done in England. Whatever he came to feel in later years about his native land, he always retained a great admiration for the sheer energy and enthusiasm shown by the design and engineering teams at Avro Canada, and never failed to contrast their approach with the sluggish attitude of their British counterparts.

By the time he and Avro Canada's own test pilot, Don Rogers, had completed the initial tests the Canuck was a much stronger aircraft. Waterton was posted back to England and Gloster sent out 'Zura' Zurakowski as his replacement. 'Zura' continued to help refine the aircraft until it was ready for production. It was now well on the way to becoming the RCAF's mainstay night fighter, the last examples of which were decommissioned only in late 1981. Zurakowski was to demonstrate the CF-100's flexibility most memorably at the 1955 Farnborough air show with a typical display of aerobatics ending in a show-stopping 'falling leaf' that brought amplified tannoy gasps echoing around the airfield.

Having spent fifteen months away in Canada instead of the expected six, Bill Waterton arrived back in Gloucestershire to find 160 new but untested Meteors parked at Moreton Valence and a backlog of administrative work that ought never to have been allowed to build up. It wasn't as if he was the only test pilot in Gloster's employ, nor does it seem credible that in his absence abroad the company wouldn't have arranged for someone else to have taken over his paperwork. It all contributed to his feeling that there was something fundamentally wrong with the way the company was organised. The suspicion was further strengthened by his next bone of contention, which was over pay. This was now a live issue throughout the industry and not merely at Gloster. Three years after his appointment as the company's chief test pilot he was still on his original salary of £1,500. A semi-formal table that compared pay scales across several of the main aircraft companies revealed that Vickers paid far above the rest, that de Havilland's salaries were tolerable, and that even some smaller companies like Fairey paid over £3,000 to their chief test pilot. Gloster Aircraft and the other companies of the Hawker Siddeley Group came bottom of the list. The worst paid of all were Avro's pilots, and their chief test pilot was faced with open rebellion in the form of a strike. The issue smouldered on in the industry, but the senior pilots' stance was often weakened by their juniors taking fright and wanting to back down. A grudging ten per cent pay rise was eventually agreed, but by the end Waterton felt he had only succeeded in alienating both sides. He had vociferously faced down management on behalf of himself and his juniors, but he had also antagonised many of his younger colleagues who were scared of losing their jobs altogether by being identified as militants. It was a typical Waterton outcome. 'My popularity was pretty low all round. Many of the pilots felt I had let them down . . . while the management thought I should have been supporting the firm's line one hundred per cent. The only thing that kept me on at Gloster's was my love of aeroplanes and flying. The set-up had precious little else to offer.'[13]

Listening to Bill Waterton reminiscing on tape in 2003 at the age of eighty-seven, it is impossible to detect any note of regret, still less of grievance or bitterness. It is the remarkably young-sounding voice of a man with a razor-sharp memory for anything to do with aircraft and flying. (It is also the voice of a man who has reacquired a good deal of his Canadian accent over the last half-century.) He is often amused by a recollection and occasionally forceful in his descriptions of the unnecessarily poor working conditions he and his fellows had endured, or equally of the inept management and political decisions of the period. Yet one would guess that fifty years earlier he would have adopted precisely the same tone with Gloster's directors: stating the inequities baldly and often with a touch of acerbic humour, but without the least concession to diplomacy or the judicious smoothing of ruffled feelings. Undoubtedly, many considered him rude and difficult; but his voice – whether reported or recorded – does not at all support the idea of a cantankerous person. Rather, it is the voice of a straightforward man who, having been schooled from an early age to make black-and-white moral distinctions, became the test pilot to whom matters in his profession, as in an aircraft, were either right or wrong. The cockpits of the early jets he flew were no arenas for compromise. Like the engineering of the aircraft itself, controls and instruments either worked properly or they could kill you well before you had digested your last piece of breakfast toast. This was likely to add a certain bluntness to the way you dealt with the world, no matter how much you disguised it beneath laconic RAF slang. 'A bit of a prang' one morning could still lead directly to having to identify the disgustingly contorted or charred remains of your friend after lunch.

No doubt this somewhat technical way of treating things overlapped into Waterton's way of treating people, and with poor results. On all evidence he was much liked and respected by friends and certain colleagues; but a suspicion remains that he became not so much *dis*liked by Gloster's management as slightly feared. Whatever anyone privately thought about feisty Bill Waterton, they couldn't easily dismiss what he was saying; and his sayings were frequently hard. His single-minded absorption in his work, combined with lengthy trips abroad (neither of which was atypical of many test pilots) must also have had repercussions for his emotional life. At the age of thirty-five he was still single, despite a reputation for being quite a ladies' man. There were surely moments when, returning to bleak bachelor digs after a day spent struggling with a potentially lethal aircraft or its makers, he felt himself beleaguered and isolated – even downright lonely. This was hardly the middle-class comfort David Lean was to depict a year later in *The Sound Barrier*, with test pilots living in oak-beamed period cottages with their toddlers and meals-on-the-table English-rose wives. But if Bill Waterton ever felt this, he never betrayed it.

1951 found him, therefore, back at Gloster with abraded feelings on both sides. Such was the background to his earliest tests of the company's newest and most secret fighter, the Javelin. It was also the start of a turbulent five years that would end with his acquiring a status within the British aircraft industry as Public Enemy Number One.

5

Canberras, Hunters and Patriotism

Another of the pilots who had left the RAF to join the Gloster Aircraft Company was Roland Beamont. A high-profile war hero for having downed so many incoming V-1s with his Hawker Tempest, Beamont acquired his first experience of jet aircraft when he flew a Meteor in 1944. Next year at Gloster he helped with the test flying of the Meteor Mk IV in which Group Captain 'Willie' Wilson broke the world air-speed record at Herne Bay. He then had a brief stint at Hatfield with de Havilland as a demonstration pilot, flying a Vampire in the first post-war SBAC show at Radlett. He made a sensational impression by throwing the little fighter about the sky with unprecedented panache. As a remarkable pilot he would anyway have been a notable acquisition for Gloster, but inevitably he lacked Bill Waterton's more recent glamour as a member of the prestigious High Speed Flight. This had made Waterton a prime target for Gloster's headhunters, just as his colleague Neville Duke had been for Hawker's. Beamont must soon have realised he was unlikely to become Gloster's chief test pilot in the near future, and he was ambitious. Within months, English Electric offered him exactly that top job, so in 1947 he left to go to Preston. He had first made contact with this company while at de Havilland, who had had to farm out the manufacture of some of their Vampires to English Electric to keep up with a flourishing export order book.

It turned out to be an inspired move on Beamont's part, although at the time it must have looked to many like career suicide. English Electric, a major manufacturer of electrical equipment as well as of trams, buses and diesel locomotives, had not designed an aircraft since the Kingston flying boat in 1924. Although it had acted as a 'shadow factory' during the war, building bombers for Handley Page (first Hampdens, then Halifaxes), its own aircraft division had been closed for over twenty years. After the war the company decided to go back to designing aircraft again and appointed a new chief designer, W. E. W. 'Teddy' Petter. Petter's family not only made engines under their own name but had founded Westland Aircraft, where 'Teddy' had already designed three aircraft including the Lysander, a rugged, high-winged light utility plane that gave sterling service during the war and had been built in considerable numbers. After the war Westland decided to concentrate on helicopters and Petter was lured to English Electric to reopen their aircraft division. When Beamont arrived at their design offices in Preston to be interviewed for the job of chief test pilot he found a wooden mock-up of their first aircraft, the A.1 twin-jet light bomber, ready and waiting. This was the future Canberra. His appointment marked the continuation of his run of luck (he had already had the good fortune to survive the war). Two years later it would fall to Beamont to test the Canberra from the beginning; but before that he was sent to the United States to acquire experience in flying jet aircraft larger than the Meteor, particularly bombers. As we know, his luck held with the once-in-a-lifetime opportunity to become the first Briton to fly faster than sound, which he achieved at Muroc in the only flight he was allowed in the new F-86 Sabre. Returning to the UK, he took the Canberra up for its maiden flight in 1949 and immediately recognised it as a quite exceptional aircraft.

The Attlee government's short-sighted conviction that neither the RAF nor the navy would need new fighters much before 1957 was emphatically not shared by the respective service chiefs or the Ministry of Defence. Nor was it by the industry, whose untidy abundance of private companies had never really paused in their output of novel designs, only a tiny fraction of which were destined to be built and an even tinier fraction of that number to become production aircraft. No matter: the government was paying. What was more, the companies knew that governments came and went, usually in far less time than it took to get a design off the drawing board and into the air. They also had the lucrative foreign market to aim for. Since Britain was in the forefront of the new jet technology, the industry was perfectly placed to supply many of the world's air forces as they began shedding their old conventional warplanes. The English Electric Co., eager to make a good impression on its return to designing aircraft, had responded to a 1945 Air Ministry specification for a light bomber to replace de Havilland's brilliantly successful but ageing Mosquito. Like the original Mosquitos, Petter's projected bomber was unarmed, relying for its own protection on sheer speed and altitude and making for great savings in weight. Against stiff competition, the design won.

The Canberra, like the Meteor, was a twin-jet straight-winged aircraft, but there the resemblance ended. Not only was it much bigger than the fighter but proportionally it had an altogether lower, sleeker profile. The wings were roughly diamond-shaped and *big*: exactly what was needed for dealing with the thin atmosphere of high altitude. In fact, the Canberra's wing area was nearly three times that of the Meteor's. Beamont first took the prototype, VN799, up from Warton's runway 26 on Friday, 13 May 1949. Despite the date, the day could hardly have been luckier. 'I very soon became aware that this was no humdrum conventional aeroplane with merely increased performance over its predecessors, but an altogether new and different experience ... It virtually flew itself.'[1] Over the next thirty-six flights some minor problems surfaced and were dealt with: an overbalanced rudder; modifications to the elevators; a simple fairing added to cure turbulence behind the cockpit canopy that had led to a slight 'snaking' from side to side. But by August it was clear the new bomber was fit to be flown in public at Farnborough in September. It was outstandingly manoeuvrable at any altitude, thanks to good reserves of power and very low wing loading. ('Wing loading' is calculated as the laden weight of the aircraft divided by the area of the wing. Thus a large wing area relative to the aircraft's mass will produce more lift at a given speed. An aircraft with low wing loading will be able to land and take off at a lower speed than one with high wing loading, and will also be able to turn faster.) Beamont was determined to make an indelible impression with the Canberra.

This he certainly did. In fact, his display went down in

the air show's annals because no one had ever seen a bomber flown like it. This was only the second Farnborough show after the SBAC had moved their displays from Radlett and there was plenty of competition for the spectators' attention. The weather was hot and clear and the string of new British prototypes being shown off included the very first public appearance of de Havilland's new Comet. John Cunningham flew it gracefully, its highly polished bare metal finish sparkling in the sunshine. There can have been few in the crowd who were not moved by its elegance and the sense that they were watching a historic occasion on which the future of air travel was being premiered. The Canberra, meanwhile, was rolling along the taxiway. It looked splendid. Spectators accustomed to seeing wartime air-craft in camouflage or drab colours were enchanted that this bomber had been painted a high-gloss sky blue with a cursive 'Canberra' in white script on the port side of the nose. (It had been given the name in recognition that Australia was to be its first buyer.) Something about the aircraft promised to be different, and it was. Beamont took off, made some immediate steep turns and came back low along the runway at full power before pulling up vertically into a half-loop. That was merely the start of an amazing six-minute sequence of aerobatics performed so tightly that the aircraft was almost never outside the aerodrome perimeter. It left the crowds 'gasping' (as one newspaper had it). They gasped still more when Beamont opened the bomb doors for a low pass and things were seen to fall out. In the 'coal hole' in the nose the navigator Dave Walker announced laconically, 'My instrumentation has gone.' 'So have my starboard engine instruments,' agreed Beamont, and at that moment the control tower radioed: 'Canberra, you're dropping pieces.' Beamont turned and descended cautiously, lining up for a single-engine landing but then finding that the starboard engine was fine and only its instruments were dead. They made a smooth touchdown and taxied out of sight of the crowds before stopping to investigate. There were wires trailing from the bomb bay and the test instrumentation pack that had been mounted there was missing. No harm done.

The display had caused a sensation and next morning Beamont was hauled in front of the display control committee and told to 'tone down' his flying from now on. He asked what that meant exactly and when they failed to be specific he inquired if they thought it had been dangerous. They said no, not *dangerous* exactly, but, well, just tone it down. It was the same reaction that was accorded Roly Falk a few years later when he rolled the Vulcan. In the minds of the committee there was evidently a right and a wrong way of flying a bomber, and the right way was *sedately*. But sedate doesn't necessarily win orders. Twenty-three years later Bob Hotz, who edited *Aviation Week*, the leading US aviation journal, reminisced:

Biggest military surprise of the 1949 SBAC show was the English Electric Co.'s sky-blue Canberra jet bomber . . . The 15,000 lb thrust from the two axial Avons made the Canberra behave in spectacular fashion . . . [The pilot] whipped the bomber, designed to carry a 10,000 lb bomb load, around on the deck like a fighter, flying it through a series of slow rolls, high-speed turns and remarkable rates of climb . . . This was the first of countless demonstrations that sold it to 15 other air forces, including the USAF, and racked up $240 million in sales with the cash register still ringing.[2]

The Canberra was the first genuinely great British post-war aircraft. It is not necessarily one of the qualifications for greatness that the United States should buy it for its own forces (although it helps). The Canberra, which was duly built under licence for the USAF as the Martin B-57, shared this distinction with only the Harrier jump-jet and BAE's Hawk trainer. One qualification for greatness in an aircraft is its ability to be constantly updated and modified for different roles and to be effective in all of them. This inherent flexibility the Canberra had in abundance. The B.Mk 2 Canberra entered RAF service in May 1951 and eventually went through twenty-seven versions, being exported to fifteen countries. In fact, English Electric couldn't keep up with the demand and the task of building Canberras had to be farmed out to Handley Page and Short Brothers, who turned them out under licence. The basic design's success was duly reflected in its longevity. India retired its last squadron of Canberras in 2007 after half a century, and it would be no surprise to hear that a Canberra is still occasionally pressed into service somewhere in the world even today. In 2006, sixty years after 'Teddy' Petter had first sketched the design, the RAF still had the odd Canberra flying: its only aircraft ever to serve continuously for over half a century. We schoolboy plane-spotters also remember that remarkable aircraft as being regularly in the news in the fifties for its record-setting: in 1951 the first non-stop unrefuelled Atlantic crossing by a jet; in 1952 the first double Atlantic crossing by a jet; in 1953, 1955 and 1957 new altitude records, including one of over 70,000 ft.

Still, back in 1951 when the Canberra entered squadron service with the RAF not everyone shared Beamont's view that it was almost beyond criticism, even taking into account that he was employed by English Electric as their chief test pilot – a role that inevitably overflowed into that of chief salesman as well as of chief demonstrator. Like the Meteor it had an engine out on each wing and similarly suffered from problems of dangerous asymmetry if one failed. Worse, the early version had an unreliable electric motor that governed the trim tabs on the tailplane. These were small tabs fitted into the elevators themselves that made it possible for the pilot to make tiny alterations to the aircraft's attitude – to 'trim' it for level flight (for example) in a way that would have been difficult using the coarser control of the whole elevator. Unfortunately, the early Canberra's trim motor would sometimes run away unstoppably, deflecting the tab to its limit and sending the aircraft nose-down in a vicious bunt that proved beyond the strength of pilots to correct. The ensuing vertical dive accounted for several fatal crashes before it was rectified. But in those days that sort of technical problem was expected with a new aircraft.

Less excusable were the defects that seemed merely thoughtless and added unnecessarily to the crew's stress. Pilots like 'Robby' Robinson, who quickly graduated from the Meteor to the much larger and more potent Canberra, loved its stability and the powerful performance afforded by its twin Rolls-Royce Avons but were outspoken about its discomforts. Certainly Ralph

Swift was when he was posted to 527 Squadron at RAF Watton in Norfolk. 'I was not designed by nature for the Canberra. I could never see adequately out of the thing. Even with the seat raised to its maximum height I was doomed to be flying mostly by instruments or looking sideways out of the cockpit bubble. The constant stretching to try and see over the cowling resulted in a very painful backache.'[3] Robinson endorsed this view. 'The aircraft was a lady,' he readily conceded, 'but the view out of the canopy was awful... I found the seat height lever and raised it until my head touched the canopy. That was marginally better, but the restricted view over the glare shield was one aspect of the aircraft that I hated throughout my association with it over the next twenty-five years.'[4] Even when you could see out, the Canberra could sometimes offer a very rough ride. The PR. [Photo Reconnaissance] Mk 3 version was designed for high-altitude photography, which was destined to be one of the Canberra's most valued roles, not least in the hands of Australian pilots during the Vietnam War. At low altitude, however, it could be a bitch. 'I learnt that high-speed, low-level photography in a Canberra was exactly like riding a bicycle with flat tyres over never-ending and badly set cobblestones,' one squadron leader was to remember.[5] The reason for this quirk was simple: the Canberra's big wings were rigid. Aircraft larger than the Canberra anyway tended to be more comfortable at low level, but something like the Victor bomber was all the more so because it had long, comparatively narrow wings that flexed and could absorb the shocks of lumpy air.

Just as wearing were the potentially life-threatening discomforts symptomatic of the rough-and-ready attitude that seemed to infect so much British aviation technology of the period. It was partly a hangover from wartime expediency, but apparently now coupled with the equally home-grown maxim that runs, 'If a thing's worth doing, it's worth doing cheaply.'

The cabin pressurisation and heating was awful. Even with heavy flying clothing and lining our boots with brown paper we froze. In the daytime, under the clear canopy, the top half of the pilot's body could swelter and, in the cockpit shadow, his flying suit would be covered in frost. With all those clothes on, and strapped tightly on to the ejector seat, it was almost impossible to reach some of the switches. The designers of British aircraft at that time seemed to ignore the pilot's comfort or even his ability to operate efficiently. Cockpits in the Fifties were ergonomic slums, with instruments and switches scattered where they fell, and the Canberra was a prime example. A long rubber bag with a chromium-plated receptacle was provided for 'relief'. With all those clothes and straps and no autopilot you had to be desperate to attempt to use it. However, we were young and knew no better. To us the Canberra was the latest and best.[6]

Evidently equipment standards had been allowed to drop rather drastically since pre-war days. The late Sir Peter Masefield, ex-chairman of the British Airports Authority, once recalled a job he had been given at Fairey in the mid-thirties as a junior draughtsman, designing a holder for a brass handbell and an ultrasonic whistle to fit into the cockpit of the Fleet Air Arm's Fairey Seal bombers. 'The handbell was to be rung after alighting in fog. The whistle was to frighten off flamingos and similar hazards before taking off from (for example) the Nile at Khartoum.'[7] In those days they thought of everything. No aids were too much for *their* pilots. It would be nice to think that Fairey's thoughtfulness was the origin of the modern businessman's phrase 'bells and whistles' to describe top-of-the-range gadgetry, instead of the more likely fairground organs.

To a schoolboy on the ground in the early fifties, of course, these defects in our jet fighters were invisible, and in any case military secrecy was such that few civilians could ever have learned about them. Adults like my parents still automatically used the wartime phrase 'hush-hush' with a slightly lowered voice to denote a conversational no-go area, which had the natural effect of making me want to know more. I had to learn from what I could see and hear, which mostly meant at air shows. Since mainly British aircraft were exhibited (exclusively so at Farnborough), these inevitably had about them a self-congratulatory, even chauvinistic atmosphere. After a day spent being battered by noise and thrilled by sights it was easy to forget that we were not the only nation on earth building aircraft, and that our erstwhile lead was even then being eroded and overtaken abroad. As it was, we plane-spotters were hard pushed to keep up with our own country's output. Our *I-Spy Aircraft* books were soon a mass of pencil ticks and we had to start proper log books to keep up with the bewildering assortment of aircraft we saw.

Those early fifties air shows belong to an era when people yearned for spectacle and entertainment after the wartime years and their drawn-out aftermath. TV sets were still uncommon and the viewing choice severely limited to evening programmes of unremitting blandness transmitted in black and white and ending with the National Anthem at 10.30 p.m. In one's memory, air displays such as those at Farnborough and Biggin Hill tend to run into each other in a stream of excited images. The hot smell of canvas in the exhibition tents and the ammoniac stench of primitive toilet facilities were the incense behind the commentators' voices over the scratchy tannoy system, now suddenly loud and then multi-tracked as the echoes bounced off hangars, often drowned completely by aero engines or the double thunderclap of a supersonic low pass. 'If you l-look towards the *east*, ladies and gentlemen, you'll see B-Bill P-Pegg-egg arriving in the new Type 170 Bristol Freighter-ter with the specially lengthened fuselage for Silver City-ity Airways' cross-Channel car-ferry service-ervice-ervice...' Meanwhile there were fairground treats on offer at a few festive stalls: Wall's ice creams that came in flattened yellowish cylinders like Price's night lights with bands of paper around them, all the more welcome since sugar was still rationed. Candyfloss, however, was banned in my medical family because of the polio scare, it being deemed even more risky than unwashed fruit. But anyway, sixpence a week pocket money didn't go far and needed to be hoarded for Jetex fuse and other vital aero-modelling necessities. '... And of course we are all waiting for the arrival of Neville Duke in the Hawker Hunter, but while we do so Tom Brooke-Smith will be bringing the Short S.B.5 over-over-over...'

I think it was at Biggin Hill that we were treated to the

most sensational new aerobatic stunt of the day: the so-called 'Zurabatic cartwheel'. Jan Zurakowski had invented it while working for Gloster. No doubt while he was testing the endless stream of new Meteors at the behest of Bill Waterton, and possibly under the goad of slight boredom, 'Zura' did something that had assuredly occurred to nobody else on the grounds that it was madly dangerous. This was deliberately to exploit the Meteor's potentially lethal asymmetric behaviour on one engine. The manoeuvre he invented required the pilot to zoom his aircraft into a precisely vertical climb to 4,000 ft on both engines until it had slowed to about 80 mph, standing on its tail. Then he would chop one throttle, causing the aircraft to pinwheel under the remaining power of the outside engine. As the nose swung downwards towards six o'clock he would then chop *that* throttle too, allowing the Meteor's rotational momentum to carry its nose back up to the vertical again, by which time the air speed was zero. The nose would fall around once more towards six o'clock, the Meteor by then having spun through one and a half revolutions while virtually stationary in the sky, and all in the vertical plane. From the spectator's point of view it was truly astounding to watch an aircraft apparently stop in the sky and pinwheel. It was a spectacular display of control, made all the more impressive because it is very difficult for a pilot to keep an aircraft exactly perpendicular to the ground while on his back, staring up into the sky. Nor did we realise how much skill lay in recovering from the manoeuvre because once it was finished the aircraft had effectively stalled, and if the nose went even slightly beyond the vertical it would flick into an inverted spin, and no Meteor had ever been known to recover from one of those, no matter who was flying it.

Zurakowski first publicly performed his cartwheel in 1951 in a Meteor F.Mk 8 fully loaded with rockets at the Farnborough air show, provoking an announcer's involuntary exclamation 'Impossible!' over the loud-speaker system. 'Zura' became famous overnight – a fame whose lustre was only increased when within weeks five Meteors were lost through ordinary RAF pilots trying to imitate him. A year or two later, though, several had mastered the technique and it was probably one of these 'Zura' imitators rather than the man himself whom we watched cartwheel a Meteor in the summer-blue skies above Biggin Hill. As already mentioned, 'The Great Zura' later joined Avro Canada to complete the testing of the CF-100 Canuck that Waterton had begun, surviving several near-death experiences when canopies failed and fuel lines ruptured. He then went on to test the Avro CF-105 Arrow. This was the fabulously advanced delta fighter whose abrupt cancellation in 1959 by government decree on 'Black Friday' is as much a part of Canadian aviation enthusiasts' collective memory as the identical fate of the TSR.2 in 1965 was to become part of Britain's. 'Zura' retired from test flying in 1958 in fulfilment of a promise to his wife Anna. He died in Ontario in 2004 at the age of 89. No one who saw his flying ever forgot it.

Details blur after more than half a century but I retain a clear, generic *airfield* memory of that period. It includes the exhilarating scent of burnt kerosene, the glitter of the aircraft on static display, the dance of thermals over tarmac and hangar roofs, the lashing grass's sheen as it was blown flat by prop-wash, and the background roar of jet engines that came in waves, intensifying and then shutting off suddenly as though a door had closed. I remember that the hangars as well as many of the brick admin buildings still wore fading stripes of wartime camouflage paint. And I remember a Lincoln bomber overflying the crowd so low that its slipstream and wing-tip vortices dragged hats and programmes into the air in a swirl as the raving bellow of the four Merlins only thirty feet away brought us out in gooseflesh. I hadn't seen it coming; and when suddenly this monstrous noise and huge dark shape pounced from behind my hair bristled and I went cold. Long afterwards I wondered whether this reaction was due to an atavistic memory dating back millions of years when some ancestral creature had emerged from the sea and tried to scuttle up an exposed beach before a winged, predatory shadow overtook it from behind.

Occasionally I come across a photograph of air-show crowds all those years ago. I turn the thin, already slightly brittle pages of *Flight International* or *The Aeroplane* and think how oddly innocent we all looked. Not just our obedience in standing so demurely behind a single sagging rope, which was often all that stopped us from walking a few yards onto a live taxiway or runway. There was something equally trusting in the artless austerity of our clothes: men still wearing their demob coats or old service raincoats, boys in shorts and Clarks sandals or plimsolls with the occasional Brownie box camera, faces thin and eager and all upturned as though we had been promised a sign from the heavens. We had, of course; and it came, too. The deafening waves of sound, the sparkling array of fabulous machinery that droned or sped past and over us, rolling and banking and blasting upwards: these were home-grown *patriotic* marvels we were shown. Nobody could have walked back to his coach with its smell of sun-baked Rexine upholstery and packed lunches without feeling proud and reassured. Elsewhere, cynical journalists might (and did) cast doubts on all this expenditure, and pin-striped men in Westminster seemed eager to cancel everything at a stroke even as worried service chiefs tried to match tomorrow's needs with today's intelligence reports of what the Soviets were doing. But we the air-show public were largely oblivious. What we wanted was spectacle, and it was given us in heaping measure. We went home knowing that Britain was still a world leader in technology and feeling in our bones our place in the international order. The sheer gusto and energy of aviation in those days felt not only right, but endless. It didn't matter that ordinary life was austere, that motorcycles with sidecars were almost as numerous as cars in the muddy parking areas. Our political docility was as much a part of being British as were the displays of military and commercial power we were treated to.

And who could be cynical for long when the proof of our superiority was so regularly proclaimed in news of yet another record broken, whether of speed or altitude or distance, or else of an aviation first? My contemporaries and I grew up to these announcements and took them for granted. In 1945 and 1946 the Meteor broke the world air-speed record. In 1946 Martin-Baker made the first reliable ejector seat. Britain broke the 100 km closed-circuit record in 1947 and again in 1948. When a Boulton Paul Balliol trainer was fitted with an Armstrong Siddeley Mamba engine it became the world's first single-engined turboprop aircraft in 1948, the same year that the Fairey Gyrodyne gained the world speed record for helicopters. When the prototype de Havilland Comet airliner began flying in 1949 it was the world's first jet airliner; and since it could fly twice as high and twice as fast as any other commercial airliner it began to break records in a constant stream. In 1950 Supermarine's Type 510 became the first swept-wing jet to land on an aircraft carrier, and that year British European Airways inaugurated the world's first-ever scheduled helicopter service. A couple of years later the de Havilland 110 became the first two-seat, twin-engined aircraft ever to exceed Mach 1. But far more impressive in late August 1952 was the Canberra's double Atlantic crossing from Aldergrove to Gander and back in ten hours and three minutes (including a two-hour stop in Newfoundland). VX185, piloted by Roland Beamont himself, became the first aircraft ever to make the double crossing in a day. The ecstatic British press quoted *The New York Times* saying that it marked a new triumph for the Canberra. 'The British jet plane industry appears to be assuming the lead in the Western World in both military and commercial aircraft – and it is a comfort that it is on our side.'[8] The Canberra's point-to-point flight records alone were soon legion, but the ones I remember were its altitude records. In 1953 Walter Gibb took it to 63,668 ft and two years later flew the same aircraft to 65,890 ft. In 1957 a rocket-assisted Canberra achieved 70,310 ft.

As for absolute speed records, they became a little thinner on the ground as American aircraft took the lead. But on 7 September 1953 Neville Duke broke the F-86 Sabre's existing world air-speed record with a run of 727 mph in an all-red Hawker Hunter off Littlehampton. Later that same month Mike Lithgow raised it to 736 mph in a Supermarine Swift in Libya – a record that lasted all of two days before a Douglas Skyray took it back for the US at 753 mph (itself beaten within the month by two miles per hour by the then accident-prone F-100 Super Sabre). It was a competitive year, and all mixed up in my memory with the climbing of Everest and the Coronation. The last air-speed record I remember is the one gained by Peter Twiss in 1956 flying the beautiful Fairey Delta 2 over Chichester. I heard the announcement with my parents on the radio news at teatime one Sunday. 'Golly!' I remember my mother exclaiming, the teapot frozen in her hand. 'One *thousand*, one hundred and thirty-two miles an hour!' When she was born in 1908, Blériot had yet to fly the Channel.

Meanwhile, behind these fifties scenes some classic British aircraft were taking shape on drawing boards up and down the country, some of which even flew after a long wait. A certain amount of what the air-show public saw was perhaps misleading, in that it consisted of prototypes that never went into production (like Hawker's P.1081 or Supermarine's Type 529 with the butterfly tail), or else they were purely experimental aircraft designed for nothing more than testing a particular configuration (like Avro's little 707 deltas mentioned in Chapter 1). In any well-run air show the appearances of these often futuristic and even odd-looking aircraft were mixed in with displays by familiar old favourites and by less dramatic transport, civil and light aircraft. And in 1953 we all saw newspaper pictures of a peculiar thrust-measuring rig that was soon universally known as the 'Flying Bedstead', the first step on the long road that led to Vertical Take-Off and Landing (VTOL) aircraft like the Short S.C.1, the Hawker P.1127 and ultimately to the Harrier itself. The overall impression was of infinite variety and fertility and it was all lent further spice by the secrecy issue. 'We're not allowed to tell you just what the top speed is,' the commentators' voices told us confidentially from a thousand loud-speakers, 'as I'm sure you'll understand-and.' Then some rakish flake of metal blasted past and our hearts soared as one. In our boyhood bibles such as the *Observer's Book of Aircraft* or my own favourite, Green and Pollinger's *The Aircraft of the World* (1953), it was equally titillating to find, amid the usual data for a particular aircraft, the italic phrase *No details available for publication*.

Over half a century later I stare at this phrase in my worn copy of Green and Pollinger as it hides a Hawker Hunter's weight and all the Avro Vulcan's performance figures, still as dutifully obedient to an ancient pact of secrecy as my grandmother was to the one she had sworn in the First World War. Even today in a certain mood I find a ghost of that former secrecy can still hang a faint aura of mystique around these aircraft, now so venerable that appeals have to be made to save the last airworthy specimens and their rare appearances still guarantee large turnouts at air shows. I well remember the first photograph of the Handley Page Victor bomber in *The Times* that my father cut out and sent me at school. It was a shot of it parked, taken from the side because, as the caption noted, the exact shape of its wings was still a closely guarded secret (they were merely progressively less swept, in a style soon known as 'crescent'). How this official secret was supposed to be kept from the hundreds of ordinary citizens or Russian spies who were bound to see it whenever the Victor flew in and out of Handley Page's airfield, God alone knew. In any case the Russians had themselves tried a similar planform for their MiG-17 fighter in 1950, so the concept was already somewhat old hat.

But anyone interested in aircraft in those days necessarily came up against a wall of secrecy. Of all the major democracies Britain has always been the least open, which is presumably why unnatural emphasis is laid on the hallowed national platitude 'freedom of speech' rather than on freedom of information. I always assumed this had to do with the country's peculiar caste system whereby 'the powers that be' retain those powers precisely by not allowing the unempowered to know what is going on. Even today, 'security' is still the magic catch-all behind which the powers protect themselves from the horrid implications of real democracy, just as it was during the depths of the Cold War.

If only it weren't so bafflingly inconsistent. One can see why it was a good idea to keep details of the Vulcan bomber's performance from the Soviets in 1953 – even though from its shape, size and known limits of engine technology any competent aerodynamicist or even plane-spotter could have supplied pretty accurate guesses. But why *was* the news of Roland Beamont's supersonic flight in the USA in 1948 withheld? And if 'national security' is to be the sacred consideration, then what on earth is to be made of that extraordinary request by Rolls-Royce in 1947 that they be allowed to sell Derwent and Nene jet engines to the Russians, despite warnings by eminent engineers like Rod Banks that it would be equivalent to 'selling our birthright'? Yet the new President of the Board of Trade, Sir Stafford Cripps, expressly gave Rolls-Royce the go-ahead. The great Russian aircraft designer Alexander Yakovlev recorded that when someone in the Kremlin hopefully suggested trying to make this purchase Stalin had called it 'naive', adding, 'What kind of fools would sell their secrets?'[9] Answer: Rolls-Royce and Stafford Cripps, aka the powers that be, who will do anything for money. So it was that the German engineers working for Soviet Russia after the war could produce no jet engine powerful enough, and the Klimov-built Nene duly became the heart of the highly able and successful MiG-15. It can hardly have endeared Rolls-Royce to American pilots over Korea.

Yet despite the official secrecy shrouding so many of the aviation industry's activities, and despite all the frantic activity visible at air shows, the fact was that by the early fifties the RAF could see a horrid gap opening up in Britain's air defences. The first-generation jet fighters in squadron service – the Meteor and the Vampire – were already beginning to show their age. Neither could have held its own in aerial combat against the F-86 Sabre, let alone against a really well-flown MiG-15. The outbreak of the Korean War had caused near-panic among British service chiefs. Thanks to the Labour government's lack of a modern aerial defence policy the RAF was without a competitive front-line interceptor. In 1952 it was obliged to shop abroad and import 428 Canadian-built Sabre Mk 4s that served with eleven squadrons until 1956, by which time they had been replaced with home-grown Hawker Hunters. This was something of a national humiliation, given the size of Britain's sprawling and still unreconstructed aero industry. All those companies, yet the lead already lost. Even so, those Sabres were by no means a case of *faute de mieux*. The F-86 was a great aircraft, especially in its Canadian-built version; and since we had so stupidly dallied there was not a hope that we could have produced anything to match it in the requisite time. Towards the end of a long career Eric 'Winkle' Brown summed up the Sabre quite simply as 'the finest jet aircraft I ever flew, from a handling standpoint. The Spitfire of the jet age.'[10]

The real problem was that successive governments failed to give any clear lead. They never managed to thrash out with the often warring service chiefs exactly what Britain's military commitments to NATO and the remaining Empire would require in terms of aircraft, and then to fund and support them. This failure was all the more pronounced against the background of those semi-private aviation companies who were already working with supersonic wind tunnels and thinking many years ahead. In the best of these there was no lack of radical ideas, and many designs were astonishingly forward-looking. Several should have become great aircraft; and if only they had been properly backed from the first by steady funding the whole post-war story of Britain's aero industry might have been very different. But so many of the projects were, like Miles's M.52, destined to be capriciously scrapped at the last moment. Many others were to be handicapped from the start by delays that meant that by the time the aircraft had been fully tested and debugged and in production they had missed their opportunity for real commercial success.

Some, like the Supermarine Swift, simply never lived up to their promise: a common enough occurrence in aircraft industries everywhere. For all that, a Swift did briefly take the official world air-speed record in 1953. The FIA ruling on speed records now specified that they had to be the best of four three-kilometre runs below 100 metres, so Mike Lithgow took the Swift to the Libyan desert where the high temperature made the air less dense. By such means it scraped into the record books for two days. Subsequently, the aircraft was never a success in RAF service, being too full of faults that were solved and sorted only after it was already obsolete. It was all deeply ironic, given its starring role in *The Sound Barrier* as the Prometheus, when Lithgow himself flew the prototype Swift in impressive and deafening high-speed passes for David Lean's camera to suggest a future of unstoppable British technological progress (*Ad astra! To the stars!*). Unfortunately, even in those days a fighter aircraft that air forces wanted to buy and fly needed to have a lot more than the mere ability to make fast low passes in a straight line. It had to have combat manoeuvrability, for one thing. By now level-flight supersonic capability was also pretty much a prerequisite; yet no British aircraft had it until Fairey's experimental delta in 1956, and the RAF had to wait for their first truly supersonic fighter until 1960, when deliveries of the Lightning began. Britain had fallen that far behind.

While watching the infinite variety at those fifties air shows, some adult taxpayers might briefly have reflected on the infinite amounts of money this sort of hit-and-miss approach to aviation was costing the nation. And this without any of them being privy to the AIB's crash records or having glimpsed those heaps of broken Meteors lurking out of sight in squadron dumps. They probably couldn't have guessed quite how much the monthly wastage in accidents alone was costing. But then it might not have mattered to them any more than it did to me. How could such dreary accounting be matched against the tingling expectation aroused by hearing the windswept tannoys announcing 'Neville Duke-uke is just approaching from the west-est in the Hawker Hunter' while we all tensed ourselves for the bang?

Certain test pilots, such as Roland Beamont, were associated with several different aircraft. Others, no matter how many types they flew, became identified with only one. In the early fifties Neville Duke's public linking with the Hawker Hunter was absolute, cemented on 7 September 1953, the year of the Coronation, when he took the world air-speed record away from America with his run of 727 mph in the red Hunter WB188. He had nearly come to grief a week earlier on his first attempt. 'One undercarriage leg was sucked out with a big bang as I ran in over Bognor Pier, passing at about 300 ft with the speed building up to 700 mph. The Hunter whipped over the vertical and I was nearly into the sea. Such was the strength of the Hunter that it held together *in extremis*. I landed her back at Dunsfold on two wheels without too much further damage.'[11] (What a way to earn a living, we may think.) Neville Duke DSO, DFC**, AFC had had a spectacularly successful war as a fighter pilot in both Europe and North Africa, with twenty-eight confirmed combat 'kills'. Once out of uniform he had been associated with Hawker's new jet fighter from its beginnings in the late forties and took the first prototype – painted a pretty duck-egg green – up on its maiden flight in July 1951. So well did it fly and so few were the details needing attention that it was ready for the Farnborough air show in September, where it made a great impression. Here at last was the proper swept-wing jet fighter the RAF badly needed and had ordered straight off the drawing board. Unfortunately, it was a further three years before the Hunter began to enter squadron service. This was partly owing to a familiar British dilatoriness, but mainly due to the discovery that when it was finally armed and flown as a fighter rather than as a prototype, all sorts of snags arose that needed to be dealt with.

Still, from the first people were ravished by the aircraft's looks. In a world of military aviation still dominated by clunky old Meteors, the Hunter seemed like an envoy from the future. Now at last the RAF could have a jet fighter that *looked* fast. The mid-mounted wings were swept back at a thirty-five-degree angle and the fin and tailplanes were also swept. Behind the cockpit's bubble canopy a low fairing continued along the aircraft's spine, blending into the fin's upwardly curving line. The single Avon's intakes were narrow triangles at the wings' roots. Somehow, this made all the difference to the Hunter's superbly trim and elegant lines. The North American F-86 Sabre (which had held the speed record Duke broke) was also neat and clean, but its jet intake was a hole in the nose: a common feature of the time, shared by the somewhat similar-looking MiG-15 and giving both aircraft a blunt, stubby appearance. The Sabre was perpetually open-mouthed like a bemused fish, which definitely detracted from its looks. (The English Electric Lightning, then still some years in the future and also a mouth-breather, had a large intake with a shock cone for slowing down the airstream feeding its two engines. It always looked as though it was choking on an ice cream cornet.) The Hunter, by contrast, could debatably be regarded as perhaps the most graceful jet fighter of all time. It was designed by Sydney Camm, who had also designed the Hurricane, and he had unquestionably produced an aircraft that was lovely to look at. Certainly I lavished more time and care on my balsa-wood model of it than I did on any other, for once making the boring effort of sanding the wings enough to get the right varying thickness or chord. I finished it in the duck-egg green of the first two prototypes and quite wish I still had it. I blew it up one night, as boys do, in an experiment involving cordite filched from a .303 cartridge, Jetex fuse and underwing pods of dope thinners. At the time the spectacle was well worth it.

The Hunter looked so good and flew so well it was perhaps easy to overlook that it could only go supersonic in a shallow dive. However, the aerodynamic problems surrounding compressibility had by then been largely solved or were being overcome, and it would go past Mach

1 without fuss or cockpit drama. 'You can take up the Hunter any time you like and put it through the sound barrier, knowing that it will perform perfectly,' Duke enthused.[12] The sonic bangs it was called upon to produce at air shows from 1952 onwards became the Hunter's expected party piece, like Barney Gumble's explosive belches in *The Simpsons*. In his autobiography, originally published in 1953, Neville Duke was rhapsodic about flying the aircraft that had become so firmly identified with him:

One of the thrills of flying is to take up the Hunter to over 40,000 feet, up into the clear, deep sapphire blue . . . Now . . . at full throttle you half-roll over and pull through. The nose of the Hunter is pointing straight down at the earth and you are hanging forward in the straps, feeling as though you may slip out of them and fall forward at any moment. Now you are really beginning to move. The indicated speed begins to build up and so does the Mach number. Soon you are going straight down at the earth at supersonic speed. You can see the earth rushing up towards you. The needle on the altimeter is whirling madly round, reeling off thousands of feet as you go down, straight as an arrow. It's a wonderful thrill. When the Hunter is going down flat out you are falling at much more than 50,000 feet a minute . . . For me there is no greater satisfaction than sitting in the cockpit of the Hunter, beautiful in design and construction, representing the thought and skill of so many people, and feeling it respond to the slightest movement of your fingers. It lives and is obedient to your slightest wish. You have the sky to play in – a great limitless expanse.[13]

Ironically, it was a Hunter that put an end to Duke's career as a test pilot. One day in August 1955 he was scheduled for some gun-firing tests off the Sussex coast. He had just taken off from Tangmere in WT562 and was at about 1,000 ft with Chichester harbour directly beneath him when he suddenly found he was only getting idling thrust from the engine. Nothing he could do made any difference. At that altitude and with no power to speak of a pilot is left with very few options. One of them is to bale out, but Duke noticed that RAF Thorney Island was coming up below him and he decided to try and get the aircraft down. As he came up in a high-speed glide his dilemma was that he was travelling too fast for a safe touchdown but not fast enough to go around and land on the main runway. He was out of options and realised he would have to put down diagonally across the airfield at about 200 mph, and on rough grass at that. As soon as his wheels touched the turf the Hunter leapt into a series of bounces that became higher and higher. In order to avoid a stall off the top of one of these, which would probably have been fatal, Duke selected 'wheels up' but a single green light remained on, showing that one leg was still locked down. WT562 crashed to the ground lopsidedly and began swerving madly.

Sitting helplessly in the cockpit, I jettisoned the hood and cut the fuel while the aeroplane careered into a number of arcs. I had no control over my destiny whilst being shaken unmercifully. Reaching the edge of the aerodrome it hurtled across a ditch before crunching nose first into a sea-wall on the other side. The machine broke up but I emerged from the mess with only cuts and bruises but aching badly. I had in fact fractured my spine.[14]

After recuperation Duke found his flying was unimpaired but that any appreciable *g* force produced excruciating pain. So that was that where test piloting jets was concerned.

The convention is to say that the Hunter was 'a joy' to fly – as indeed it could be. But the fact is that it wasn't much of a joy at the beginning, being plagued by several faults, some of which were cured over time. The least curable was that it had 'short legs', in the parlance. By nursing his fuel a pilot could keep the Hunter usefully in the air for about an hour, but for precious little longer. (The same problem, only much more acute, was to dog the Lightning.) The incident of six out of a flight of eight Hunters running out of fuel simultaneously over East Anglia has already been alluded to (p. 125). That was merely an extreme case of a not uncommon event. When it finally entered restricted squadron service in 1954 the Hunter did so with firm limits set on its speed and manoeuvres. One such limited 'manoeuvre' was the quite basic one for an interceptor of firing its guns. It had been found that the Avon engine was anyway prone to surging owing to the design of the intakes. ('Surging' was a sudden reversal of the flow inside the compressor – a sort of jet engine backfire. It often resulted in a flame-out, or complete loss of power.) A redesign partially cured this but it was then discovered that when the 30 mm Aden cannon were fired their ingested gases could also trigger surging and flame-out in the engine. A good number of Hunters and their pilots were lost in this way. What was more, the ammunition belt links that were ejected when the cannon were fired would cause considerable damage to the underside of the fuselage until the problem was cured by collecting the links in a couple of bulgy fairings that became known as Sabrinas in honour of Norma Sykes, an outrageously pneumatic starlet of the day. (In some squadrons the aircraft themselves were known as Sabrinas.) Another of the Hunter's foibles was its tendency to stall in high-g combat turns: not an ideal characteristic for a fighter. Over the years fixes were found for most of these glitches, but it all meant that the Hunter was yet another example of a British military aircraft that had been put into service before it was fully sorted. A giveaway is that Neville Duke should still have been carrying out cannon-firing tests when the aircraft had already been in squadron service for a year. Defects like that ought to have been fixed before ordinary pilots were expected to fly it. It was akin to marketing a car that stalled every time its fog lamps were turned on. A telling indication of how slow the Hunter was in becoming a fully capable warplane is that even in 1956, when Anthony Eden foolishly committed Britain to the invasion of Suez with Operation Musketeer in October, the AIB crash records show Hunters dropping from the sky like shot pigeons all over the UK as pilots of ordinary skill were faulted either by its demands, by fuel shortage or by mechanical failure of one sort or another.

Only in 1957, with the newly engined Hunter F.Mk 6 entering squadron service, did the RAF finally have an outstandingly capable fighter. Pilots found it a great improvement over their Canadair Sabres, which is hardly surprising given that the F-86 had been flying since 1948. In due course it became Britain's second most successful post-war fighter aircraft in commercial terms after the old

Gloster Meteor. Nearly 2,000 Hunters were built and the type served for decades in nineteen air forces around the world. It became one of the RAF's mainstays until the end of the sixties, seeing action in Aden and Indonesia. Even before the fifties were out the F.Mk 6 became particularly famous as the choice of RAF aerobatic teams like the Black Arrows and, later, the Blue Diamonds. People lucky enough to have been at the 1958 Farnborough air show saw the Black Arrows' sixteen all-black Hunters joined by six more F.Mk 6s from other units to perform a spectacular double loop of twenty-two aircraft, a record that still stands.

There is a school of thought that maintains the Hunter was never developed to its full potential, becoming blighted – like so much else in British aviation – by Duncan Sandys's 1957 White Paper that mistakenly foresaw missiles taking over from fighter aircraft in the immediate future. It is a debatable viewpoint. But the fact remains that even by 1956, two years after the Lightning's prototype had flown, the barely transonic Hunter was already an old generation fighter. The Lightning itself went into RAF service in 1960 and was capable of exceeding Mach 2, while no amount of upgrading could ever have made a Hunter fly much faster than Mach 1, not even with steeper sweepback and an uprated engine. Nevertheless, it ably proved itself in Aden during the Radfan crisis of 1964, flying low-level ground attack sorties against Yemeni rebels. The Hunter became much loved in the RAF and also by the general public. It looked like the sort of competent warplane a country could be proud of and even non-enthusiasts could appreciate its aesthetic appeal, just as they could later with Concorde. It was also immensely popular in overseas air forces, where it frequently saw action and was prized for its toughness. By 2008 the Hunter had been in service with the Lebanese Air Force for fifty years, and there are at least fifty privately owned Hunters still flying around the world. Neville Duke, who died in April 2007, would not have been at all surprised to be outlived by the aircraft he had long ago made his own.

6

Crash Landings

Back in 1948 the Gloster Aircraft Company had responded to an Air Ministry requirement, made the previous year, for a high-performance night fighter. The company's chief designer at Bentham, George Carter, must have been a professional of great versatility as well as one who moved with the times. Before the war he and Henry Folland had jointly designed the Gladiator and, following a visit by Frank Whittle to Gloster, he had then designed the first British jet aircraft, the E.28/39, around Whittle's engine. He followed that immediately with drawings for the Meteor. After the war Carter was much influenced when the various radical new wing shapes that German scientists had been experimenting with came to light. He was particularly struck by the possibilities offered by the delta.

As professors Busemann and Lippisch had already determined, the delta wing looked as though it could solve some tricky transonic aerodynamic problems by delaying compressibility while at the same time providing plenty of lift, especially at altitude – not to mention offering extra space for fuel and weapons. So in 1948 Carter submitted a series of designs for the night fighter the Air Ministry wanted. The one chosen was a delta known as the GA.5: the future Javelin. By late 1951 the prototype was ready for

One of the three High Speed Flight Gloster Meteor F.Mk IVs practising near Bognor Regis for the successful attempt on the world air speed record in July 1946. The pilot is probably Bill Waterton. In the background the two barrage balloons, Gertie and Ermintrude, wait to be deployed as course markers.

De Havilland's DH.108 'Swallow'. TG283 was the first of three prototypes, each designed for testing at different speeds. This was the aircraft that Geoffrey de Havilland Jr first flew from Woodbridge in May 1946. All three types were destined to crash and kill their pilots. In 1950 TG283 spun into the ground near Hartley Wintney.

A line-up of 54 Squadron's brand new de Havilland Vampire F.1s on the day they were delivered to RAF Odiham, 29 June 1948. In the following month these were to become the first jet aircraft ever to fly across the Atlantic (although not non-stop).

1950. The wartime night-fighter ace 'Cat's Eyes' John Cunningham is now de Havilland's chief test pilot. He is standing in front of the prototype DH.106 Comet 1 at the company's Hatfield works. The world's first jet airliner, which he has just been testing, is already in BOAC's 'Speedbird' livery as G-ALVG.

The blue-painted prototype of the English Electric Canberra B.1, VN799. The pilot is Roland Beamont, the date 23 August 1949.

An Avro Canada CF-100 Mk 5. In January 1950 Bill Waterton flew to Canada to help with the testing of the prototype 'Canuck'.

The Supermarine 535 Swift prototype, VV119, flown by Mike Lithgow: the very aircraft that appeared as 'Prometheus' in David Lean's 1952 film *The Sound Barrier*.

A navalised version of the DH.110 that crashed so spectacularly at Farnborough in 1952, the Sea Vixen gave good service in the Fleet Air Arm. XP924 was originally an FAW.2 and served with 899 NAS before being retired. Now beautifully restored to airworthy condition, the type's last flying example is pictured here at Biggin Hill in June 2009.

The Gloster GA.5 Javelin: the DH.110's rival for the same RAF requirement. WD804 was the prototype, seen here flown by Bill Waterton in early 1952. He had first flown it on 26 November 1951, since when it exhibited a wide range of handling problems that Gloster's design team downplayed in defiance of their chief test pilot's warnings.

The end of WD804, and very nearly that of Bill Waterton as well. The aircraft lies in a puddle of fire-fighting foam after its epic crash at Boscombe Down on 29 June 1952. It can clearly be seen that both elevators are missing after the Javelin shed them in mid-air following catastrophic 'flutter'.

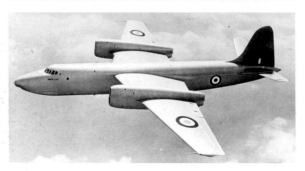

A Hawker Hunter F.Mk 1, WT557. The type was surely one of the most beautiful jet fighters ever.

Fairey's chief test pilot, Peter Twiss, puts the FD.2 into a climb in March 1956. The little Fairey Delta was the last fixed-wing British aircraft to capture a world air speed record.

Short's S.A.4 Sperrin (seen here in August 1951) was built to the same 1946 specification as that of the three V-bombers, but to a more conventional design.

The Vickers Valiant B.1, the first of Britain's V-bombers to go into squadron service, shows off its graceful lines. Britain's first atomic and hydrogen bombs were dropped by Valiants during tests in 1956 and 1957. This particular aircraft is WZ365, which had ten years' service with RAF Bomber Command before being broken up for scrap in 1965.

An Avro Vulcan B.2 performing its celebrated leap into the air on take-off: most un-bomberlike behaviour that never failed to thrill airshow spectators like these at Farnborough in September 1959.

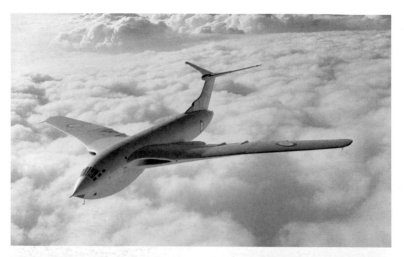

Handley Page's chief test pilot, S/L H.G. Hazelden, flying the HP.80 Victor prototype in July 1953. The Victor was to play a crucial role as a tanker in both the Falklands War and the Gulf War. The aircraft illustrated is WB771 which, together with four crewmen, was lost on 14 July 1954 when the tail fin sheared off in flight.

A diamond formation of English Electric Lightning F.6s of No. 74 Squadron, based at Leuchars. The type's phenomenal performance was allied with hamperingly short endurance. Without finding a tanker, a Lightning on a typical sortie had a bare 40 minutes' flying time.

A Blackburn Buccaneer S.1 (top) joins an S.2 in typical low-level flight, most likely in 1965. The Fleet Air Arm's Buccaneer was immensely robust and, although subsonic, was widely regarded as perhaps the finest low-level strike aircraft of its day.

The ill-fated English Electric/BAC TSR.2. This is prototype XR219, flown by Roland Beamont and Don Bowen in the autumn of 1964, photographed from Jimmy Dell's Lightning T.4 flying 'chase'. The TSR.2's shock cancellation by the Labour Government in 1965 remains bitterly controversial to this day.

On 29 July 1950 a British European Airways' Vickers Viscount 630 prototype inaugurated the world's first scheduled commercial passenger service by a turbine-powered aircraft. It is shown here leaving Northolt for Paris with fourteen fare-paying passengers aboard plus Sir Peter Masefield (then head of BEA), the Viscount's designer George Edwards and the jet engine pioneer Sir Frank Whittle.

G-ANBC, a Bristol Britannia 102 in BOAC livery on 5 July 1955. A potentially great long-range passenger aircraft when it was designed in the late 1940s, the Britannia's development was fatally delayed by slack management at Bristol and disgraceful vacillation by BOAC.

Roll-out day at Weybridge for the first of BOAC's Vickers VC10s, 15 April 1962. At the time the VC10 was the most powerful airliner in the world and, but for the grotesque delays that had come to characterise the British aircraft industry, might have proved a commercially world-beating design.

Gloster's chief test pilot caught in the late 1940s in characteristic pose. Bill Waterton's humour and professionalism won him loyal friends in the aviation fraternity, while his outspokenness and intransigence earned him some dedicated opposition from within Britain's aircraft industry in the early 1950s.

The old aviator shortly before his death: Bill Waterton at 90, a portrait by his elder son Willy. Appearances might suggest that the previous half-century had not greatly softened the former critic of Britain's early Jet Age. Friends, however, remember him fondly for his sardonic wit and unfailing civility.

The Hawker Harrier's predecessor, the P.1127 FGA.1 Kestrel. Fl/Lt Fred Trowern demonstrates a vertical take-off for a group of VIPs at Bircham Newton on 30 September 1965. The Harrier became the world's first and (for decades) only successful VTOL aircraft. It joined English Electric's Canberra as one of the rare British military aircraft to be bought and built under licence by the United States.

testing and it was Bill Waterton's job to do it. By then he must have been looking forward eagerly to flying the futuristic new aircraft after years of bread-and-butter testing so many production Meteors. Relations between certain of the company's management and their highly skilled but frustratingly intransigent employee had already been strained by Waterton's critical approach to technical matters such as cockpit layouts, not to mention his campaign for better pay. Gloster's management now left him in no doubt that they expected him to polish off the Javelin's exacting test programme in the shortest possible time. Everyone at the company was very conscious that de Havilland had just produced a much larger swept-wing development of their successful twin-boom Vampire and Venom fighters, the DH.110, which was also designed to be a high-performance night fighter. It and the Javelin were therefore two completely different outcomes of the same 1947 Air Ministry requirement and were viewed in both companies as rivals competing for the RAF's approval and production contract. The Gloster board noted with particular anxiety that John Derry had already been flying the DH.110 for two months over at Hatfield before the Javelin could be rolled out. It was a lead Waterton was expected to wear down and then take himself.

Of the two aircraft the Javelin was by far the more radical in design: a great sixteen-ton brute with a large delta wing and a huge slab of fin topped in a 'T' shape with a smaller delta tailplane. Its twin Armstrong Siddeley Sapphire engines were intended to boost it quickly to a height of 50,000 ft, where it would supposedly intercept high-altitude Soviet intruders. It was the world's first twin-engined delta aircraft and also the first delta with a tailplane. Bill Waterton had been familiar with the concept since George Carter first sketched it out three years previously, and he privately viewed it as Carter's masterpiece that would crown the designer's distinguished career. But even as the wooden mock-up was being built he had raised a pilot's objections to certain design details, such as the cockpit's restricted view. Far worse was the Javelin's proposed control system, which was identical to that of the cancelled Gloster 'Gormless' and meant the controls would be so heavy at speed as to tax the limits of a pilot's strength: a ludicrous handicap for someone expected to concentrate on using his aircraft as a weapons platform in air-to-air combat. But as so often before, Waterton's pleas for change before it was too late went unheeded. Apparently it was *already* too late. The faults were faithfully incorporated into the Javelin prototype so that when it first flew, as he himself recorded, 'I was the least surprised of anyone that its controls were hopeless. They had no positivity, and were virtually immovable at more than half speed, even with two hands.'[1]

The taxiing trials and early flights revealed other problems, though no more so than in any new aircraft. Waterton recorded that basically the Javelin was easy to fly, had potentially excellent performance and showed great promise, though it also had some dangerous tendencies that needed correcting. However, when he reported these, 'only lukewarm interest was shown'. By now George Carter had retired from designing and was on Gloster's board of directors. His former staff under the new chief designer, Richard Walker, treated the Javelin as their baby, 'and like doting parents, they blinded themselves to its faults – and took most unkindly to anything which might reflect on their parenting'. How amazed and uncomprehending we schoolboy plane-spotters would have been had we known the absurdities that went on behind the scenes in the test flying of one of Britain's top-secret aircraft! Not long after the Javelin's first flight Waterton was asked to put on a demonstration to impress some visiting men in suits from the Ministry of Supply. He gave them a display artfully designed to look impressive while enabling him to keep the speed low. He performed no loops, either then or at the two Farnborough air shows at which he was subsequently to demonstrate the Javelin, because he knew he would never have the strength to pull out of a steep dive. At the end of his demonstration a senior Gloster man from the Bentham design office remarked acidly that he didn't know why Waterton made such a fuss about the controls if he could throw the aircraft about like that.

This was all the more enraging in its revelation of how little some designers appeared to understand about flying. It also brought home to him how radically attitudes in the industry had changed in the last few years. During the war test pilots had customarily worked in league with oil-stained engineers and hands-on designers who, when there was a problem, would eagerly rush away and come back with some inspired 'fix', whether it was an ingeniously machined widget or a redesigned undercarriage mechanism. By contrast, it now appeared that the industry of the new Jet Age was run by accountants and boffins, none of whom ever flew in their firms' creations and, in

Waterton's experience, always showed the greatest aversion to doing so. The accountants wanted maximum profit in the shortest time and the boffins were happy producing 'concept' brochures with ever more elaborate graphs and tables. Neither seemed much interested in the end users of their products. As Waterton was to reminisce later: 'At the Central Fighter Establishment a *flying* man ran the show. He said what he wanted, and got it. But in the industry, the men who fly the planes seemingly came second to schedules; and the chief test pilot at the end of the line was a necessary nuisance who must not be allowed to interfere in the practicalities of commercial enterprise.'[2] For Waterton, who was still instinctively on the side of the average RAF pilot, this depressing new regime was made still worse by the chronic dawdling and lack of urgency that seemed to define the entire enterprise, even despite the need to overtake de Havilland's DH.110. Nor was his mood helped by memories of his recent experiences testing the CF-100 in Canada, where problems had been instantly addressed with the utmost energy and enthusiasm.

Things came to a head in April 1952 when he handed in a letter of resignation. This must have badly shocked Gloster's management since they immediately rejected it with a show of great affability, assuring him that he was invaluable to the company and from now on would be listened to and well looked after. Reluctantly, Waterton decided to give the company a further chance. The decision nearly cost him his life. On 29 June he took the prototype Javelin, WD804, up for yet another test flight, its eighty-fifth. It was a Sunday and there was now a real urgency in his test schedule. The following day Fighter

Command was holding its annual tactical conference at which he was due to demonstrate the Javelin and John Derry the rival DH.110. By now Waterton had already taken on the DH.110 in mock combat at 35,000 ft and believed the Javelin, with its low wing-loading, to be far superior in manoeuvrability. He was keen to show the RAF what a remarkable aircraft it really was. So now he put it into a high-speed practice run at 3,000 ft over Oxfordshire, heading roughly in the direction of Brize Norton and Witney.

Then, without warning, WD804 turned itself into a sort of crazy pneumatic drill. Landscape, instruments – everything – fuzzed into a blur. My scrambled brain registered one horrifying word: 'flutter' . . . Before I had time to close the throttles or open the air brakes there were two explosive cracks. Then an uncanny, ominous silence as the rattling ceased. The nose pointed itself downward towards the ground which was only seconds away. Something had let go, something in the vital, pitch-controlling elevator circuit – perhaps the elevators themselves. I had a vision of the whole tail having broken off. The stick confirmed my assessment: at that speed normally almost solid, it offered no resistance to forward or backward movement.

My immediate impulse was to get out. Yet I knew that no one had baled out at such a high speed and lived to tell of it. So I restrained myself and considered the faint possibility of getting higher and slower. There were one or two things worth trying. I could alter the plane's flight path by varying engine power – perhaps by flaps and air brakes too, when speed was low. On the other hand any of these actions might upset the plane's equilibrium. But I had to do something. With one hand on the canopy jettison handle and the stick between my knees to keep the plane flying level laterally, I gently inched the trim wheel back with my left hand. It worked. The nose rose. We started to climb, gradually the speed fell off, and I eased the throttles back gently. At 10,000 feet I levelled out doing 300 knots. Now getting out was a practical proposition.

I switched my radio to 'transmit' and spoke to my controller, Roy Julyan, at Moreton Valence. 'I've had a spot of bother, Roy. No elevator control. The stick is no longer connected to anything fore and aft. The elevators went between Brize Norton and Witney. I've got up to ten thousand and I'm now heading for the Bristol Channel where I propose to dump the lot in the drink.'

That, indeed, was my plan, for no one with a conscience could abandon several tons of explosive fuel and metal where it might fall on a town. But I could not stay up for ever. The engines were thirsty brutes. Yet as I flew westwards I began to toy with the idea of achieving the seemingly impossible – of *landing* the Javelin. At 10,000 feet there is room to play about, so I explored my makeshift handling of the plane at circuit and landing speeds. I found I was able to fly with undercarriage down and flaps partly lowered, as for landing. I also found that I could keep the Javelin to within two hundred feet of any height I wished, and that if I were to drop too low I could put myself in a climb by applying power. Why not have a try at landing the Javelin at Boscombe Down on Salisbury Plain? My primitive instinct of self-preservation was urging me: *Get out, you fool; get out of the plane*, while my reason was saying: *You must try to save the aircraft*.

At last Boscombe Down hove in sight. I called the tower and asked the controller to get his fire tender ready. With wheels and part flap left down, I gradually went down to fifteen hundred feet, keeping my speed around the 200 mark. A wide circuit with the gentlest of turns, and I was aligned heading west towards the main runway. Now to reduce height and speed. Holding her laterally level with the stick between my knees to prevent turning over, my left hand was on the trimming wheel, my right stretched across my body to work the throttle. Now I was down to a thousand feet . . . now down to five hundred, with my speed 60 mph fast, for below this I found she lost steadiness and what little control I had. I would have to put her down at two-thirds above her normal touchdown speed. At this speed she would fly herself off the ground without help from the pilot; but if I could catch her on the ground with the wheel brakes I might pitch her forward and keep her down at the small risk of blown tyres and burned-out brakes. The aerodrome boundary passed and she went in as steady as one could wish. If only a gust of wind didn't hit us. The runway loomed up. As I eased back the throttles she touched down with but the slightest bump – I had pulled off a 'daisy cutter'. Yet a split second later all was lost.

Whether it was a fickle variation of wind, or an undulation in the runway, I shall never know. The Javelin, still very much a flying machine, bounded into the air again to drop gently back on to the runway – to be flung aloft again by her tough, springy undercarriage, aided by her great buoyant wings. In a succession of ever-increasing bunny-hops we bounced along the runway, higher and slower, higher and slower – with a heavier bang every time we grounded. Onlookers later said the Javelin could easily have cleared a hangar. Inside, it was rough and frightening, sitting waiting for something to go. It soon did.

We dropped from a near-stall with an almighty crash. This was it – a dull, heavy boom, the smell of paraffin, and a sheet of flame and black smoke slashed over the cockpit. Fuel tanks had ruptured and were exploding. This time she stayed down for a bit. The port leg, breaking away from its mountings, had been driven up through the wing. The plane lurched over, her port wing-tip dragging along the concrete, swinging the Javelin off the runway. I was thrown to the right of the cockpit despite my straps – then to the left as we were flung crabwise into the air again with momentary smoothness. Then another crumpling noise as she came down on her nose and rocked to her starboard wing-tip, slithering sideways and collapsing that undercarriage leg. Another dull boom as the starboard tanks went up. More grinding of metal signified a rapid spin. A stagger as we settled into a dust-spewing heap – then silence, except for the terrifying roar of flaming paraffin.

So far so good: now to get out. In order not to disturb the airflow over the broken tail I had left the canopy closed. It had served me well when the tanks went up, saving me from the spray of burning paraffin. But now it was to be my jailer. Flames roared fiercely behind, to each side and over me. I shut off the fuel cocks to the engines, burning my knuckles on the hot side of the cockpit. I pressed the button which operated the electric motor to push back the cumbersome and heavy metal and Perspex canopy. It did not budge.

A macabre picture flashed across my mind. I saw again the torso I had found in the ashes of a crashed Harvard. There was no head: a three-inch strip of what seemed like leather was all that remained of the neck. Twisted black rosettes represented the shoulder sockets. Streaks of darkened blood revealed one thigh joint. My gloved hand lifted a charred waistband to disclose human skin. Above everything the smell: the composite odours of flesh and oil, human fat and charred metal . . . The picture came and went. The hood remained jammed shut. The heat was suffocating and the sides of the cockpit made me wince when I touched them. I turned the regulator to 'Emergency' to breathe pure oxygen in the acrid, smoke-filled cockpit. I groped for the crowbar to lever open the canopy or break the plastic. It was not in its spring clips on the port side of the fuselage. It had shaken itself free, and was later found under the seat. I had no time to find it and again cursed all designers as the Perspex alongside my head began to melt and sag inwards.

I banged around the cockpit like a man gone mad. I cursed, pressed buttons, pulled, tugged and heaved – but nothing would yield. Neither the jettison handle nor the canopy would give a fraction of an inch. A new thought – the ejector seat. I was sitting on a miniature artillery shell capable of hurling the seat and me some sixty feet in the air through the canopy. What if the heat at the bottom of the cockpit should set it off?

My luck held. At last, persistent attempts at the actuating button combined with banging the canopy with my fist, head and forearms took effect. It gave a fraction, enough for me to get my fingers under its front arch. The sheer brute strength of desperation helped force it two-thirds open. In a flash I was out and put fifty yards between myself and the blazing wreck.[3]

Once the fire tenders had arrived Waterton grabbed one of the crews' hoses and directed its plume of foam on to the Javelin's nose where the aircraft's all-important recording instruments were: the only bits of the wreck worth saving. Eventually the fire was brought sufficiently under control for him to retrieve them. An RAF doctor bandaged his scorched arms and anointed his singed eyebrows before he defied medical orders, borrowed some clothes to replace his own foam-soaked overalls, and scrounged a lift back to Moreton Valence. The Javelin's elevators were later found in a field near Witney, victims of violent 'flutter'. Not only would there be no Javelin to display to Fighter Command on the Monday but it was clear the type was still far from ready to be passed as suitable for production. The crash and Waterton's narrow escape were headline news.

For his skill and determination in bringing the aircraft back instead of abandoning it over the Bristol Channel Waterton was awarded the George Medal. The local Gloucestershire newspaper printed a brief 'interview' with him, a cutting of which he kept. '"Bill's" reaction to the honour is typically modest. "I rather look upon this as a recognition for test pilots as a whole, and not particularly myself," he told *The Citizen* today.'4 Beside this piece of high-minded fiction 'Bill' pithily scribbled 'Balls'. The George Medal's inaugurator, King George VI, had intended that it should be awarded sparingly 'for acts of great bravery'. In due course Waterton the Canadian received it with appropriate humility from the hands of the new Queen Elizabeth. Waterton the Gloster employee, on the other hand, was reliably less humble. He later wrote that some people would have been happier had he abandoned the aircraft. 'Certain boffins and designers never quite forgave me for bringing back the bent plane as conclusive evidence.'5

After the accident Waterton decided he still could not resign from Gloster because he was worried people might think the experience had 'finished' him (as he put it) and besides, he hated to leave a job half-done. Ironically, that summer of 1952 was when John Derry had thoughts about leaving de Havilland and joining Gloster. The two men met but Derry understandably balked at the smaller salary Waterton was authorised to offer him. It would be surprising if Waterton had not also warned his fellow test pilot that other things at Gloster were not entirely rosy. At any rate nothing came of the plan. Waterton's log book shows that on 28 August he flew Derry from Brockworth to Moreton Valence in a de Havilland Beaver, so this may have been the occasion. That must have been a busy day because later that afternoon Waterton made a flight in a Comet 2, G-ALYT, as second pilot with John Cunningham, presumably after he had returned Derry to Hatfield.

Nine days later John Derry was dead, along with Tony Richards and twenty-nine Farnborough spectators. Even as ambulances were still rushing to the scene Bill Waterton dutifully climbed into the second Javelin prototype as Neville Duke took off to ensure the show went on, then followed Duke's Hunter into the air, sad for his friend and considerably anxious for himself lest the defect that had so recently wrecked WD804 should recur. Not only did the show go on but his aerobatics in the Javelin, restricted though he kept them, must have looked to many as proof of the aircraft's superiority over its de Havilland rival, now scattered in fragments all over the airfield. However, Waterton knew all too well that the Javelin might just as easily come apart in mid-air as the DH.110 had, albeit for quite different aerodynamic reasons. The vicious flutter that had affected WD804's elevators on that disastrous flight only two months earlier had shaken the entire aircraft enough to blur his eyesight. At the time it had seemed to last for several minutes but it was later found to have been a mere two and a half seconds before the hinges had fatigued and the elevators snapped off. Had it continued for longer the entire aircraft could easily have broken up. In the harrowing circumstances and with his burn scars still visible it must have taken a good deal of nerve as well as professionalism for him to put on a public display that afternoon at Farnborough. His log book is probably revealing of the man in that it reveals nothing – other than that he flew the Javelin at the SBAC display on each day of Farnborough Week 1952. Not even a note marks 6 September as the day of his friend's death and the greatest air-show catastrophe to date.

Waterton was to spend nearly two more years as the Gloster Aircraft Company's chief test pilot. In view of his differences with management it might well be asked why he stayed. Thanks to his spectacular accident and award he was now a celebrity and a household name. With the immense experience that underpinned his public acclaim one would imagine he could have found another job as a test pilot without much difficulty. On the other hand the world of aviation was comparatively small and he knew his reputation in some quarters as a 'difficult' or 'bolshy' character would count against him. Besides, his reluctance to leave the job of testing the Javelin half-finished was typical of the man, who was nothing if not doggedly conscientious where his profession was concerned. All evidence suggests that even after six years as a civilian he was still heart and soul on the side of the RAF and determined to tame the Javelin before service pilots were expected to fly it. In any case, he could not leave Gloster without first finding a suitable successor. In the autumn of 1952 he chose Peter Lawrence, an ex-navy pilot who had been chief test pilot for Blackburn Aircraft up at Brough in Yorkshire.

Meanwhile, Gloster were feeling the pinch. The

government had recently struck two hundred Meteors off their order book and the Javelin was still a long way from production. The gap thus created in the firm's revenue led to redoubled pressure on Waterton to finish the testing and approve the aircraft as soon as possible. He flew the third prototype in March 1953 while the second was being fitted with a modified wing. When that one returned to Moreton Valence he made the first dozen flights in it to establish the new wing's handling characteristics before turning it over to Peter Lawrence for a second opinion. He was still not happy about the aircraft's behaviour at low speeds, finding the elevator control sluggish, but he thoroughly briefed Lawrence, who already had several hours' experience of the aircraft with its original wings. Lawrence accordingly took the aircraft up on 11 June and within the hour came news that he had crashed near Bristol and was dead. It appeared that the Javelin had experienced a deep stall. If so, it was one of the earliest instances of a then unrecognised aerodynamic phenomenon that can affect an aircraft whose tailplane is mounted T-wise on top of the fin. At a certain angle of attack the wings can prevent the airflow from reaching the tailplane, screening it off and leaving it becalmed in dead air. In such conditions the elevators become useless. At some point Lawrence's Javelin had pitched up steeply enough for its huge wing to have acted like an air brake and without any elevator control it had stalled in this stable state, dropping vertically like the flat iron it resembled. AIB investigators later noted that the grass around the crash site had not been flattened, suggesting that at the moment of impact the aircraft had had no forward momentum. Lawrence, desperately trying everything he could think of to destabilise the Javelin and regain control, must have left his ejection too late. He was found still strapped in his seat near the aircraft's burnt-out carcass. He was thirty-two. Waterton, deeply upset by the loss of a close friend and colleague he had personally appointed, at once hastened to perform the grim task of breaking the news to Lawrence's wife Barbara before she could hear about it on the radio.

The exact cause of the crash was never officially established but Waterton had his own ideas about several likely contributory factors. Three years later he expressed them in print: two inflammatory sentences in his anyway provocative book *The Quick and the Dead* that were to be seized on by Gloster and others in a furious reaction that was given prominence by the national newspapers. 'As a result of the accident,' he wrote, 'the firm altered the flaps. It was little consolation to me that it took the death of my number one to effect modifications that all my talking and reports had been unable to achieve since the aeroplane's earliest flights.'[6] No doubt at the time of the accident and in his trademark forthright manner he made it clear to his colleagues at Gloster that in his opinion the company's sloth and selective deafness had helped kill their own test pilot. However, he must have taken good care that Barbara Lawrence did not get wind of the accusation to add to her distress. When the *Daily Mail* reviewed Waterton's book in 1956 his allegation evidently came as a shock to her. She was quoted as saying, 'I knew there were faults in the Javelin at the time of Peter's crash, faults they didn't seem able to put right. But perhaps in order not to worry me Peter did not say much about it. There had been a lot of unsatisfactory flights. Bill Waterton is not the man to make charges lightly. To know the crash could have been avoided is even worse than what I went through before.'[7]

At the time Waterton's expressed view could hardly have improved his working relations. Yet not every prospect was glum because this was the year in which he married. Marjorie Wood was an ex-Wren who was a first-rate statistician as well as a former fashion model of striking good looks. They had known each other since July 1944 when Waterton had been testing weapons over the Wash and she had been in radio contact with him from the Ops. Room. 'There was this confident, very kind voice with a wonderful Canadian accent,' she later recalled. 'I told myself that I was going to marry that man some day.' Nine years later she did, on 26 July 1953. Bill was thirty-eight.

That was also the year of the Coronation, when on 2 June I and millions of other British children became New Elizabethans, exhorted to look back to the glories of the Tudor era and forward to the even greater national brilliance that the crowning of the young second Elizabeth surely presaged. Our mood was upbeat. Had not sweets just come off rationing in February, and wasn't all sugar rationing to end in September? Was not the sky full of marvellous winged portents of British ingenuity and aerial might? And didn't the *Eagle*'s centre-spreads weekly dissect the workings of such things as the Supermarine Attacker or the Bloodhound missile? Periodically, one of the *Eagle*'s trademark cutaway drawings was of some imaginary vehicle of the future that somehow managed to look both visionary and outmoded at the same time. 'The Supersonic Jet Bomber of Tomorrow' was one such: a rakish deltoid dart with six engines mounted across three rear tail fins and an 'atomic bomb bay' (number 14 in the key). Surely, we felt, the future was already within our grasp in the brains of our visionary boffins and in the skilled hands of men like Neville Duke and Bill Waterton. How could it go wrong?

Waterton finally left Gloster on 31 March 1954. The firm later made it clear that he was sacked. It has been suggested that the order most likely came from Hawker Siddeley Group chairman Sir Roy Dobson himself. 'Dobbie' had always been a notorious hirer-and-firer, quite capable in his prime of walking through a factory floor, spotting someone whose work he thought substandard and sacking him on the spot.[8] Unionised industrial practice had curbed him somewhat but no doubt the instinct lingered, especially in times of crisis, and the Gloster Aircraft Company was by now in urgent need of money. It was still waiting for the injection of government funds dependent on a firm contract for the Javelin to go into production for the RAF, and it seems probable that Dobson and the other directors viewed their chief test pilot's stubborn refusal to pass the aircraft as downright obstructive, and finally as endangering the entire company. At any rate the parting was abrupt. Waterton's log book reveals that his last flight for Gloster, when he took a Meteor F.Mk 8 up on 31 March, also marked the last day of his employ. Once back on the ground he collected his gear and was out, after seven years and five months. In that time he had played a major part in selling some 1,200 aircraft abroad, earning the company and the country many millions of pounds. He had logged 1,800 hours in the air, much of it highly demanding and dangerous flying. Even with a few modest pay increases and bonuses over the years his annual take-home pay had averaged £1,100,[9] less than an airline pilot's and with many times the risk: not a bad bargain for the Gloster Aircraft Company. Bill Waterton's celebrity was now such that his surprise departure was front-page news in the national dailies.

As for the Javelin, it was to be a further two years before it went into RAF service with a single squadron, no. 46. As with the Hawker Hunter, its speed was regulated and certain manoeuvres restricted pending further modification. In the following ten years it went through as many versions until it finally evolved into a useful all-weather fighter. It saw service during the Malaysian–Indonesian confrontation between 1963 and 1966, based in Singapore and flying patrols over Borneo. In 1964 it scored its only air-to-air combat victory: an Indonesian Air Force C-130 transport that crashed while trying to avoid a Javelin that had been sent to intercept it. But by then the design was nearly twenty years old, even if its active service life had been barely half that, and the type was finally withdrawn in 1968. It is quite usual for aircraft to be improved as they are flown, but the Javelin's degree of modification was exceptional. It had been yet one more instance of a British aircraft being hustled into service before it was fully developed and its test pilots could sign it off without lingering misgivings. Bill Waterton's conviction that the

Javelin was never ready to go into RAF service while he was flying it was amply borne out by its later history.[10] It turned out to be the last aircraft that Gloster ever built.

For me, a certain melancholy still surrounds the Javelin, partly because I associate it fondly with Bill Waterton but partly also because it did look wonderful in the air. There was something powerful and purposeful about that solid delta, and it retained a rakish air even on the ground, where it seemed to squat slightly nose-high as though waiting to leap up at the first sooty bang of a starter cartridge. (In actual fact the starters sometimes exploded, and wily Javelin pilots tended not to strap in until both engines were running.) Like the DH.110 it was a two-seater and its twin Sapphire engines punched it along in a way that didn't seem lumbering, for all its size. We New Elizabethans who saw it in flight or illustrated in magazines were overawed by its futuristic design and thrilled by the 'top secret' aura that surrounded it for so long. None of us for a moment suspected the negative nature of many of the secrets it hid. We wanted to believe the best of it, as we wanted to believe the best of our country and the industry that had created the Javelin and all the other awesome aircraft that filled the skies above Hampshire during those magical Farnborough Weeks once a year.

As for its great rival, the DH.110, it was perhaps brave of de Havilland to continue with its development after the Farnborough crash. Also, the RAF seemed to be favouring the Javelin because they thought its design was simpler and its maintenance would be cheaper. But de Havilland persevered, and when the Royal Navy expressed interest the much-strengthened 110 was navalised for carrier-borne operation as the Sea Vixen. It gave faithful service with the Fleet Air Arm in sundry crises around the world until it was phased out in 1972 and replaced by the American F-4 Phantom. Until 2017 there was still one airworthy Sea Vixen, and to see it in silhouette banking against cloud brought back a rush of memories from what felt like the dawn of the Jet Age. There was something quintessentially de Havilland in the shape of its swept-back wings, derived as they were from the third DH.108 Swallow's, and all the more noticeably so because the twin booms and tailplane made the fuselage look truncated. And then one glimpsed in it the small fleeting ghost of the Sea Venom; and suddenly a whole line of descent stretching back to a drawing board in the later years of the Second World War was visible in this quick shape against the twenty-first century sky. But these days, following a hydraulic failure and wheels-up landing at Yeovilton in May 2017, XP924 looks unlikely to fly again.

The real symbol of Britain's status as an aviation world power in this post-war period was not its scrappy attempts to build a true supersonic fighter but its V-bomber fleet. Fighters were generically impressive, especially if they were supersonic; but they remained tactical: aircraft with short range and high performance that could be scrambled to meet an incoming threat or to deter movement on the ground. Bombers, on the other hand, were clearly strategic, operating as part of an internationally agreed NATO pattern of containment. At the end of the war when Germany's remaining V-2 rockets and their research material fell into Allied hands, it was clear that in principle their technology could be developed to deliver the new atomic bomb. This meant that at some time in the future there might no longer be any need for slow, heavy aircraft to fly long and vulnerable missions deep into enemy territory in order to drop conventional bombs. Rockets could be developed to deliver atomic warheads from many thousands of miles away. The concept of the Inter-continental Ballistic Missile had been born.

However, in 1946 such a weapon was still a long way from being practicable, so plans were drawn up for a generation of bombers powered by the new jet engines. These aircraft would be very much faster than the old Lancasters and Wellingtons and Halifaxes that had given such good service during the war. Also, they would be carrying atomic weapons so vastly more powerful than ordinary high explosive that even if only one bomber got through to the target the effects would still be devastating. Washington was soon making it clear that Britain had a vital part to play in containing the Soviet Empire within its already considerable post-war boundaries, and that meant nuclear deterrence. At this point Britain's military strategists, who included the Cabinet ministers in Clement Attlee's post-war Labour government, found themselves in a dilemma. On the one hand they recognised the great desirability of close military co-operation with their American allies: had not Attlee's predecessor Winston Churchill only that March delivered his 'Iron Curtain' speech in Fulton, Missouri, after receiving an honorary degree, where he specifically referred to the 'special relationship' between Britain and the United States? On the other hand there was no question that Britain had a proud, centuries-long tradition of military and political independence to maintain.

That same year, 1946, America's focus on consolidating its new role as a superpower led to its establishment of the Strategic Air Command (motto: 'Peace is our Profession') that was intended from the outset to have global reach. Britain accepted that it, too, was required to pull its weight in containing the Soviet threat and that a new nuclear-capable bomber force was a vital part of this strategy. Still, in its increasingly paranoid manner, Washington was never wholly convinced that Britain's new Labour government wasn't somehow soft on godless communism (and admittedly the recent sale to Stalin of Rolls-Royce Derwent and Nene turbojet engines, two of the most advanced the 'Free World' had produced, can have done little to change this view). Then in August 1946 the US passed the McMahon Act expressly forbidding the sharing of its nuclear technology, even with allies like Britain. Since British scientists had made an essential contribution to the development of the first atomic bomb and it had been agreed in 1943 that the US and the UK would share such knowledge after the war, this came as a profound shock to Britain and caused great ill-feeling. The Labour government therefore determined that, though broke and heavily in debt, the country should develop its own atomic weapons, and a force of jet bombers would be the necessary delivery system for an independent nuclear deterrent.

Accordingly, in 1946 the Ministry of Defence issued OR 229, an Operational Requirement specifying a jet bomber with a range of 3,350 nautical miles (the UK to Moscow and back, with a safety margin), a maximum speed of 500 knots (575 mph, Mach 0.875) and with an over-the-target altitude capability of 50,000 ft. The bomb load, either nuclear or conventional, was to be 20,000 lb. The aircraft would carry a crew of five: two pilots, two navigators and a radio/radar countermeasures operator. Now began a curious process. Even as the designers at Vickers, Handley Page and Avro worked to produce drawings of a bomber that met these requirements, the MoD hastily issued a much lower specification for an alternative bomber as if they were suddenly worried that the first set of specifications might after all prove too difficult to achieve. Short Brothers & Harland responded to this new specification and in due course put forward their S.A.4 Sperrin, a conventional straight-winged design. This aircraft was given the official go-ahead even as the Air Staff were still trying to choose between the other submitted designs: Vickers's Valiant, Handley Page's Victor and Avro's Vulcan. At this point the 1948 Soviet blockade of Berlin suddenly goosed them into panic. Now they really *had* to have a new aircraft to equip Bomber Command, and without delay.

The upshot was that everybody was told to build everything in case one of the aircraft didn't work. In fact, this solution was not as ridiculous as it sounds. In an astonishing burst of technical creativity Vickers, Handley Page and Avro had each come up with a completely different wing shape, and all three designs were at the absolute limit of current aerodynamic knowledge and jet-engine technology. It was simply not possible to predict which would turn out the best, although Vickers's Valiant, being a shade more conventional, looked like being the safest bet. Ordering the Sperrin also represented the Air Staff playing safe in case none of the alternatives proved good enough, even though they could see its design was conservative, even slightly stodgy. In 1951 it became the first of the bombers to fly, which it did well though undramatically, and the two prototypes were destined to hang around at RAE Farnborough for some years and prove quite useful for research purposes. Meanwhile, contracts for prototypes of all three other aircraft had been signed. By now plenty of people were wondering how on earth Britain could afford to pay for the building and development of *four* different bombers merely to meet the requirements for one. They pretty soon realised there was no good answer other than that the specifications laid down by the Ministry of Defence's Operational Requirement demanded some risky pioneering with various technologies that were still a long way from being fully developed.

In all modesty, the outcome was impressive. Each of Britain's three V-bombers was radically different in design from its rivals, and not one of them looked even remotely like anything the Americans or anybody else was building. Of the big three firms, Vickers finished their Valiant first – and an outstandingly graceful aircraft it was, too. Its birth involved an entirely typical combination of serious funding and make-do-and-mend. If anybody wanted a comparison between the Heath Robinson order of the day that so often prevailed in British aviation and the money-no-object conditions enjoyed by the USAF, they needed to look no further than the final assembly and first flight of the prototype Valiant. Like all Vickers aircraft at the time it was completed at their Fox Warren plant, a hangar in some woods near Weybridge in Surrey, and then taken by road for final assembly at their test airfield at Wisley. Only a handful of miles as the crow flies, these road journeys still involved immense low-loader lorries and had to be conducted in the small hours of a Sunday morning when there was hardly any traffic about. They also necessitated the removal and replacement of all the telegraph poles along the route. Vickers's airfield at Wisley was exactly that: a field. In due course in May 1951 the first of Britain's serious nuclear bombers took off from a grass runway with Vickers's chief test pilot Joseph 'Mutt' Summers at the controls, its bicycle landing gear (one narrow wheel behind the other on each leg) leaving ruts deep enough for a man to hide in.* As a result, Vickers temporarily moved its flight operations to the British Overseas Airways Corporation's former maintenance base at Hurn, near Bournemouth, which had good hangars and a decent airstrip, while a proper concrete runway was finally laid at Wisley.

The Valiant was high winged with the inlets of its four Avons staggered at the shoulder. It was a nearly straight-winged aircraft but with some sweepback on the leading edge. The tailplanes, which were halfway up the fin, were also slightly swept. It was a lovely aircraft to look at and to fly, and 'Robby' Robinson retains fond memories of the brand-new one he and his crew were allotted in 1955 when

---

* Summers's nickname stemmed from his habit during his days in the RAF of urinating against the rear wheel of his plane prior to take-off, like a dog marking a lamp post: in his case not 'for luck', especially, but because crash-landing with a full bladder could prove fatal.

it finally went into squadron service to deliver Blue Danube, Britain's first operational nuclear weapon. It was painted anti-flash white all over as a protection against the bomb it was to drop from high altitude. 'It smelt lovely, plasticky and gluey,' he recalls. 'I liked the Valiant. I can best describe it as a big Canberra in its handling, but the really big difference was that it had a very efficient pressurisation system. When we flew at 40,000 ft plus we could relax with our oxygen masks off and in a wide, warm cockpit. No more did we have to pack our boots with brown paper and wear two pairs of gloves. It even had plumbing for one's "relief" and stowage places for one's flight bag and rations . . .'[11]

Like all three V-bombers the Valiant was intended for high-altitude, subsonic bombing; but once it was in development the Vickers team at Weybridge came up with a variant, the B.Mk 2 Pathfinder, which was expressly designed for low-level operations. Although it looked the same externally, the airframe was very different from that of the normal Valiant in order to cope with the greater stresses of low-level flying. It made its first appearance at the Farnborough SBAC show in 1953 and immediately became known as the 'Black Bomber' on account of its all-over black finish for night-time operations. It looked sensational; but then so did the ordinary Valiant in its ghostly anti-flash white. In 1956 Valiants saw action in the Suez campaign and one dropped Britain's first atomic bomb in Australia. The next year another dropped the first hydrogen bomb over Malden Island.

Alas, this elegant, handsome aircraft came to an untimely end, being overtaken by events: specifically, the shooting down over Russia in 1960 of an American Lockheed U-2 'spy plane' piloted by Gary Powers. This incident precipitated rather more than just acute embarrassment for the Americans. It revealed that sheer altitude, which hitherto both the USAF and the RAF had been relying on to keep their bombers invulnerable, was no longer protection against the USSR's latest anti-aircraft missiles. At a stroke, therefore, the Valiant had to be redeployed as a low-level attack bomber, a role for which it had not been designed. By 1964 it was discovered that the extra stresses of low flying were causing serious fatigue fractures to appear in the airframe. Indeed, at really low level the aircraft had a flying life of only twenty-five hours. As a result, the Valiant had to be withdrawn and scrapped at the end of the year after only nine years' service. The irony was that the Black Bomber variant would have been perfect for this role, but unfortunately it was now too late.

Handley Page's Victor first flew in 1952 and the Mk 1 entered RAF service in 1957. It was always the oddest-looking aircraft of the three V-bombers. Its wings seemed too far forward and its nose bulged in a strange manner. At the other end the tailplanes had a sort of flyaway aspect because they were fastened on the very top of the fin with a marked dihedral: tilted upwards from the horizontal. And then there was the famously secret crescent planform of the wings, which also had rounded tips. As a plane-spotter I could never get very enthusiastic about the Victor, although it did turn out to be a useful and reliable aircraft. What was more, it was the only one of the V-bombers that could supposedly go supersonic. As Tony Blackman (who tested the Vulcan) wrote later, he wondered what the autocratic Sir Frederick Handley Page, who died in 1962, would have thought when his firm was forced to close down in 1970, whereupon his arch-enemy Avro took over the design responsibility for Victor Mk 2s and promptly converted twenty-four of his precious bombers to tankers for in-flight refuelling to replace Valiant tankers.[12] Sir Frederick might have been slightly mollified had he known that Victor tankers were to play a crucial role in the Falklands War in 1982, and to a lesser extent in the Gulf War of 1991. The last Victor tanker was withdrawn in 1993. The aircraft was in service with the RAF for a respectable thirty-six years.

The Avro Vulcan remains by far the most charismatic of the three V-bombers – many would say the most charismatic of *all* Britain's post-war aircraft, even including the Hawker Hunter and the English Electric Lightning. In design terms, too, it was arguably the most advanced, for when it was conceived nobody fully understood all the aerodynamic possibilities and characteristics of a pure delta wing. Certainly building one with a span of 111 ft was enough of a gamble for Avro to build three third-scale, single-seat models that flew as the Avro 707 in order to test how this novel dart-like shape behaved at different speeds. The Vulcan turned out to be a brilliant design, although for many Britons its reputation was perhaps only belatedly sealed by its performance at the very end of its career in the Falklands War, which itself took place at the furthermost end of Britain's logistical capability. Quite apart from that, though, anyone who ever saw the Vulcan's prodigious kinked delta tilt across the sky, spread like a cape trailing smoke and decibels sufficient to make the earth move, and did not feel their heart leap up like Wordsworth's (who, poor fellow, had to make do with rainbows) probably had the sensibility of a cake of soap. As Roly Falk so memorably demonstrated at Farnborough air show in 1952, right from the first this magnificent aircraft could be handled like a fighter. Indeed, later Vulcans were shown to be able to outmanoeuvre McDonnell Douglas F-15 fighters in high-altitude mock dogfights.[13] (To be fair, though, it is hardly surprising with a wing that size.) In its early form, however, when it still retained its pure delta shape with straight leading edges, the Vulcan could prove a handful for the incautious to fly. This may well have been a contributing factor to what happened at Heathrow in the autumn of 1956.

On 1 October that year Vulcan XA897 was returning to the UK from Australia. Vulcans were not due to enter squadron service until the following year, leaving it to Valiants to bomb Egypt in the Suez campaign at the end of October 1956. But as the first of its kind to be accepted for service by the RAF, XA897 had been sent on a sort of 'show-the-flag' world tour which had been a great success. The aircraft had performed perfectly and had awed people everywhere as the epitome of British technological expertise and air power. In the meantime someone had taken the decision that on its triumphant return this nuclear bomber should land at London Airport (as Heathrow was then still known), where a welcoming committee of top brass and international press would be assembled to greet it. It was one of those daft notions where the urge to play to the gallery had stifled commonsensical objections such as that the Vulcan, being a military aircraft, was equipped for the GCA (ground-controlled approach) system rather than the civilian ILS (instrument landing system) with which the airport's traffic controllers were more familiar. This may have had some bearing on what was about to happen because on that day visibility was quite poor, with rain and low cloud.

Flying the Vulcan was Squadron Leader Donald 'Podge' Howard, DFC. Instead of his usual co-pilot he had the commander-in-chief of Bomber Command, Air Marshal Sir Harry Broadhurst, sitting in the right-hand seat. In the cramped compartment behind them were four men: the aircraft's three crew members from RAF Waddington – including the co-pilot – plus a support engineer from Avro. They sat strapped in as Howard brought the Vulcan in on its final approach, flared* and landed with terrific impact in a field of Brussels sprouts 600 yards short of the runway. The aircraft immediately bounced back high into the air, at which point the two pilots ejected. XA897 fell back and disintegrated in a tangled fireball of blazing fuel and wreckage that skidded along the runway. The four crewmen were killed instantly while Howard and Broadhurst were still falling beneath their parachutes. 'A thunderous explosion rent the air,' as *Time*'s correspondent put it in characteristic prose. 'In the grass alongside the runway where his ejector-parachute had dropped him,

---

* 'Flaring' refers to the pilot lifting the aircraft's nose just before touchdown. The greater angle of attack with reduced power briefly increases lift to allow a gentle landing. The same raising of the nose on take-off, but with the engines at full power, also produces more lift and is known as 'rotating'.

Pilot Howard lay, scratched and dazed but otherwise unhurt. Near by, on the concrete itself, was Sir Harry Broadhurst. His feet were broken. In a moment both airmen were in the arms of their wives who had come to cheer their return. Farther down the runway, the other greeters watched in silence as airport firemen fought the flames, and experts prepared to investigate whether mechanical or human failure had struck down the Vulcan.'[14]

In aviation circles debate has raged over this accident for over half a century. The definitive cause is still undecided and will remain so, but consensus has settled for pilot error with sundry mitigating circumstances. Among these was visibility. Even in perfect conditions the forward view from the Vulcan's cockpit was never good, and its early windscreen wipers were downright bad. It is quite likely that on that day of mist and drizzle neither pilot could see the ground at all. However, had the aircraft's official co-pilot been in the right-hand seat (instead of a brass-hat who, though a veteran pilot himself, was hardly experienced on the type), he would have been calling out the altimeter readings as was normal practice so that Howard could perhaps have decided to overshoot and go round again. Above all, it can't have helped that the wretched squadron leader must have felt under pressure, required to land a big new bomber in poor visibility for a VIP reception at a civil airport with an air marshal sitting next to him. As for the possibility of mechanical failure, *Time*'s phrase about experts preparing to investigate the crash proved optimistic because they didn't have long to examine the cooling wreckage as it hissed and spat in the rain: the airport authorities quickly sent in cutters and bulldozers to push it off the runway so that scheduled flights could resume. Evidence was almost certainly lost or obliterated. Still, what happened physically when the Vulcan's wheels had hit the ground prematurely was established. The impact smashed both undercarriage legs backwards, driving them up into the delta wing's trailing edges and severing all the control rods to the flying surfaces. From the moment the aircraft bounced back into the air it was no longer controllable, and the pilots wisely ejected.

Taking place so publicly, the accident did little for British aviation's morale, still deeply bruised by the Comet disasters two years earlier (and discussed in Chapter 7). But the heated newspaper debate that promptly broke out was over the four dead crewmen. Why had they not had ejector seats? Was it because they were just 'other ranks' and not deserving of the life-saving privileges granted to the officer class on the flight deck? In fact there *was* provision for the crew of a Vulcan to escape, but not by using ejector seats. It meant their having to put on parachutes and, regardless of *g* forces and the aircraft's attitude, drop through the entry hatch under the nose. This tortuous arrangement did save lives in a few other incidents but in this particular case could never have done so. Not only had there been no time and not enough altitude, but when the undercarriage was down even this emergency exit was hindered by the leg of the nose wheel. Public feelings ran high about the issue precisely because everybody could grasp the problem. It was easy enough to arrange for a pilot to have an ejector seat in his cockpit, but what was to be done in a bomber with several crew members stationed within the body of the aircraft? As mentioned in Chapter 4, Folland's chief test pilot 'Teddy' Tennant had successfully ejected from his stricken Gnat fighter only two months earlier. Having pondered the matter for a year while inquiries into the Vulcan crash proceeded, he wrote a letter to *Flight*:

I have read recently that Vulcans are now going into squadron service and I think it is pertinent to enquire, sir, whether all crew members now have ejector seats ... Human error is unavoidable, but in the light of evidence from the London Airport crash it would be criminal to allow our bombers to fly without equal escape facilities for all crew members, to say nothing of the ultimate effect upon morale.

To which the magazine's editor – whose own life had presumably not been saved by an ejector seat recently and whose own morale was clearly unimpaired – responded in an authoritative tone:

Only the pilots in any of the three V-bomber types have ejection seats. The crew escape system was prepared in accordance with official requirements stated some years ago ... In any case the fitting of further ejection seats would raise an almost insuperable engineering problem. We agree wholeheartedly with our correspondent's sentiment in general, though we do not feel that the effect on morale in this instance will be any more serious than it was with some marks of Meteor or the Canberra B(I).8.[15]

In the next issue of the magazine, in a note headed 'Vulcan Exit Facilities', a representative from A. V. Roe &

Co. said his company had in fact sketched out a modification that provided upward ejection for non-pilot crew members of the Vulcan but that the Air Council had decided against it after the problem had been 'exhaustively considered'. 'The structural difficulties involved in the installation of upward-firing ejection seats could not be overcome without excessive delay and without imposing unacceptable penalties in the build-up of the V-bomber force; and that downward-firing seats would be entirely unacceptable for low-altitude escape.'[16] This last was undoubtedly true: one or two contemporary American jet aircraft such as the Douglas X-3 Stiletto and Lockheed's F-104 Starfighter had experimented with downward-firing ejector seats which, when used in the case of engine failure immediately after take-off or before landing, had predictably tragic results. Even the B-52 bomber had downward-ejecting seats for its rear crew which were useless below 800 ft. However, the obvious point deliberately not being addressed was why 'official requirements' hadn't seen fit to design-in such ejector seats for all V-bomber crews from the start, when they needn't have led to delays and huge extra cost.

Not that the crews' position was always one of helpless passivity in times of crisis. As Vulcan XH498 was landing in New Zealand in October 1959, wind shear caused it to break its port undercarriage leg on touchdown. The pilot went round again for a successful crash landing. During the circuit, the crew in the rear compartment came forward and inserted the safety pins into the two pilots' ejector seats, announcing that either they would all get out together or nobody would. They all survived unscathed.[17]

The real difficulty with providing ejector seats for V-bomber crews was that since all three aircraft types were designed to fly at 50,000 ft or even higher, the flight deck and the crew's cramped, coal-hole-like compartment had to be pressurised. An ejection at that altitude would be instantly fatal. Attempts were made to design a system whereby the whole compartment could be ejected as a unit, like the one that was to be built for the pilots of the North American B-70 Valkyrie, but it proved too complex and costly. In 1960 Martin-Baker successfully modified a Valiant with rearward-facing ejector seats for the crew, but for structural reasons it would have been much harder to do the same for the Victor. Martin-Baker also designed a Vulcan crew-ejection system that worked perfectly in all twenty tests, although such seats needed to boost the men high enough and quickly enough to clear the tall tail fin and entailed severe acceleration. Had people known in, say, 1960 that V-force aircraft would be in service for a further quarter-century, it is possible that the expense of providing their crews with ejector seats would grudgingly have been met. Despite plenty of later testimony by V-bomber crews that they were generally too busy and intent on their tasks even to think about ejector seats, the fact remains that it was an awkward subject if only because from the three bombers' earliest designs it was obvious that the lives of the two pilots must have been valued above those of the crew.

Yet such concerns faded into the background as successive Vulcans slightly modified their wing shape and the true potential of this majestic aircraft was revealed. Reference has already been made to its being able to outmanoeuvre high-performance fighters at altitude, but this was only one of the abilities conferred on it by the 111 ft wingspan. The Vulcan's take-off was famous, and not only because of the characteristic moan caused by the immense torrent of air gulped through its intakes, a pulse-quickening sound that could be heard for miles. Hitherto, heavy bombers had tended to leave the ground with apparent reluctance after using all available runway. Yet as the Vulcan reached flying speed and the pilot rotated and its giant wing bit into the air it seemed to leap off the ground, climbing away as steeply as a fighter. The paradox was that the aircraft weighed over a hundred tons yet always managed to suggest an inbuilt yearning to be airborne.

Gradually, as Soviet fighters improved their capability at altitude, NATO's tactical requirements changed and the RAF's Vulcans were reconfigured to fly below any radar. The aircraft promptly revealed a low-level capability unmatched by any large bomber either before or since. By the sixties its terrain-following radar was very advanced, and now and then ground crews claimed to find leaves and even branches wedged into a Vulcan's control surfaces after a low-level sortie. RAF Vulcans would occasionally participate in Red Flag exercises: highly realistic war games staged from Nellis Air Force Base in Arizona, north-west of Las Vegas. After one such exercise a photograph was pinned to a notice board in the officers' mess. It showed the deep furrow allegedly carved in the Nevada Desert by the wing-tip of a Vulcan keen to show its

American hosts what low flying was *really* about. The Vulcan was unscathed, barring some touching-up of its camouflage paint: an extraordinary testimony to the aircraft's structural strength. It is probably a safe bet that no other military aircraft of any nationality could have survived hitting the ground while banking steeply.

But the exercises that probably did most to secure the Vulcan's reputation among NATO's military partners were kept secret from the general public until declassified in 1997, when Russia was judged to be no longer a threat. These were Operation Skyshield, three massive exercises conducted to test US and Canadian air defences against an attack by Soviet nuclear bombers. NORAD (North American Air [now Aerospace] Defense Command) was fully aware that only one bomber needed to get through in order to wipe out a city, but its planners were confident of the effectiveness of the great chain of radar stations spread across Canada and down the north-east coast of the US. This was the DEW (or Distant Early Warning) Line. To test it, NORAD invited RAF Bomber Command to participate in a Skyshield exercise in October 1961, and two squadrons each sent four of their new B.Mk 2A Vulcans. One group flew from Scotland to 'attack' from the north, coming in over Labrador; the other four took off from Bermuda and came in from the south. It is doubtful if the North American defenders realised quite what stiff opposition the Vulcan presented, carrying as it did the latest electronic countermeasures (ECM) equipment.

The four Vulcans from the north came in high at 56,000 ft. By now the airspace above the American and Canadian eastern seaboard was thick with fighters trying to detect and 'lock on' to the intruders. One Vulcan was intercepted and 'shot down' by an alert F-101 Voodoo over Goose Bay; the other three got through to land unopposed in Newfoundland, having well and truly breached the DEW Line. Three of the Vulcans coming from the south put up a joint electronic screen to jam the defenders' radar, behind which the fourth sneaked around and landed undetected in New York State, twenty minutes' flying time from New York City and a mere six minutes from Montreal. NORAD must have felt it was just as well the Soviets wouldn't be flying Vulcans. In theory, the Brits could have obliterated Washington DC, New York City and even Chicago before running out of fuel. It was deeply sobering.

Two decades were to pass before this highly versatile aircraft was finally used in anger, during which time it had been steadily overtaken by technological advance: hardly surprising for a design that dated back to the original 'V'-bomber specifications of 1946. The Vulcan had also been overtaken by strategic change. Nuclear bombers had become obsolete, their role taken over by ICBMs. In 1969, just over a year after Bomber Command and Fighter Command were merged, the V-bombers' task of delivering Britain's nuclear deterrent was finally taken over by the Royal Navy's Polaris submarines. (This ironically reversed the apparent subordination of the navy to air power that had been established in the Second World War.) Yet although it had become long in the tooth the Vulcan somehow never really seemed its age and still doesn't, even today. Maybe this is because there is something inherently ageless about big deltas: a permanently futuristic quality shared by Concorde as well as by the fabulous but aborted Mach 3-capable B-70 Valkyrie – not to mention the single most evilly beautiful aircraft of all time, Lockheed's SR-71 Blackbird.

Whether ageing or ageless, in 1982 Vulcans and Victors, the proud remnants of Britain's V-bomber fleet, were called upon to carry out Operation Black Buck in the Falklands War. This was a series of bombing raids on the airfield at Port Stanley designed to render it unfit for further use by the Argentinian air force. The Vulcans and their supporting Victor tankers were based on Ascension Island, and at the time it was claimed that the Black Buck raids were the longest-range air attacks in history. At any rate they represented the Vulcan's only 'live' combat role in thirty years' RAF service. Although since 1968 the RAF's bombers and fighters had been amalgamated as Strike Command, it is not too fanciful to view the Black Buck raids as representing in a ghostly way the swansong of the old Bomber Command. Not since 1945 had British-built bombers carried out such extreme-range sorties on a target; and they never will again. Like Bomber Command itself the Vulcan was finished, and the last few examples of this peerless aircraft were finally retired in March 1984.

Seen in retrospect, Britain's independent nuclear V-force was magnificent. The aircraft themselves were a marvel of engineering and aerodynamic design achieved on the very frontier of what was technologically feasible at the time. That we could ever have built them seems nothing short of miraculous from the perspective of the twenty-first century, when Britain's manufacturing capability has been almost entirely reduced to assembling foreign-designed cars in foreign-owned factories. Intended to absorb and then retaliate to a first Soviet attack well before the Americans could play any part, the RAF's comparatively small V-bomber force probably could never have guaranteed MAD, or Mutual Assured Destruction. But it did at least make it abundantly clear to the USSR that it risked losing five major cities even though Britain itself might be rendered uninhabitable. To that extent it was a deterrent that worked triumphantly. With hindsight, it went a long way to make up for the post-war Attlee government's complacency over the need for fighters. In fairness it has to be said that Attlee faced hideous economic problems as well as the electorate's declared priorities, which were that Labour should implement its promises of a huge housing programme while getting the new National Health Service as well as state education up and running. In the immediate aftermath of the war military matters were unpopular and being Minister of Defence was a particularly ungrateful post to hold. The ex-Vulcan flight commander Andrew Brookes sums it up neatly:

[The Labour government] maintained the atom bomb programme and work on the delivery systems [i.e. the V-bombers] through economic hell and political high water while trying valiantly to resurrect the economy, and it would be churlish and politically naïve to chide them for not having done more. In fact, when the Conservatives returned to office in November 1951, Winston Churchill, who had long criticised his predecessor's apparent neglect of the nation's nuclear defences, was amazed to find that so much groundwork had been done.[18]

In its iconic way the Avro Vulcan became one of the world's great aeroplanes, instantly recognisable. It is not surprising that it alone of the three V-bombers should have been the subject of an intense private campaign to have a last airworthy specimen fly again. After copious Heritage Lottery funding and years of dedication XH558, based at British Aviation Heritage Bruntingthorpe, near Leicester, flew again at air shows until it was finally grounded in 2016. To judge from the crowds' rapturous reception the sheer spectacle of an aircraft that looked utterly unlike anything else in Britain's skies today was, together with its sound, overwhelming. It was both an aesthetic and a nostalgic experience, exciting the young and bringing tears to the eyes of those old enough to remember the heyday of their nation's technological prowess and lament its passing. The Vulcan was not merely a dazzling monument to what Britain as a nation could once do, but a bitter reminder of what it will never do again. Whether or not we judge that this matters, few of my generation of Britons went home after a Vulcan display without pride and an empty sense of loss.

'Nothing like a hundred per cent aeroplane'

I suppose school exams must have been distracting my attention in 1954 because it took me several months to catch up with astonishing news concerning Bill Waterton. Immediately following his abrupt dismissal from the Gloster Aircraft Company at the end of March he had resurfaced in a brand-new career as the *Daily Express*'s air correspondent. Since my father took *The Times* (before switching briefly to the *Manchester Guardian*), it was some time before I realised that I could read my hero's views on aviation on a practically daily basis in the local library. His highly individual voice now gave him the personality he had lacked as a mere idol and my interest in him suddenly became more adult. If the outrageous glamour of aircraft like the Avro Vulcan, together with the heady clouds of jet exhaust at air shows, had been keeping us plane-spotters in a state of brainless optimism, Bill Waterton's articles were a vigorous corrective.

His transformation into a Fleet Street journalist was actually not quite as surprising as it might seem. The *Daily Express* was the most popular and influential British newspaper of the day. With a circulation of some four million copies it was the undisputed voice of Middle England. Its editor Arthur Christiansen was interested in science and technology and knew his readers were fascinated by anything to do with aviation. Yet behind the ritually bullish and patriotic coverage of home-grown, record-beating aeroplanes streaking about Britain's skies he recognised a growing public awareness that all was not well. This had certainly become apparent back in January when, after some previous accidents, a Comet airliner disappeared over the Mediterranean. A mere three months later, on 8 April, a second Comet became a total loss in the sea off Italy, causing the fleet's immediate grounding pending exhaustive investigation. Given the Comet's status as the world's first jet airliner, this was not only a disastrous commercial setback for BOAC but a grave blow to national morale. All of a sudden Christiansen realised there was a need to cover such aviation stories factually and without bombast. This required a writer with first-hand, up-to-date knowledge of the industry from the inside who could explain technical matters clearly and simply. With a reporter's contacts he would have heard of Waterton's reputation for crisp, even ruthless judgements about current British aircraft and the industry that was building them. When at the moment of the second Comet disaster Christiansen learned that this famous test pilot with a George Medal was suddenly out of a job, he quickly snapped him up.

It proved an excellent appointment. Over the next two years and three months Bill Waterton's frequent columns became required reading for anyone wishing to keep abreast of developments in aviation. They were also celebrated – in some quarters feared – for their remorseless critique of the manner in which Britain's military and civil aircraft were procured and produced. The revelations of a bemedalled ex-test pilot who knew the business from within could not easily be dismissed; and in this true-blue paper Christiansen occasionally felt obliged to point out to his readers that *real* patriotism required one to address deficiencies in the nation's military so that they could swiftly be made good. This was no time, he said, for adopting comforting head-in-the-sand postures. A Cold War was raging in which communism would happily exploit the least weakness in our defences . . . In the office Waterton regularly briefed his colleague Chapman Pincher on aviation matters. Pincher was the *Express*'s science and defence correspondent and Waterton later described him as 'a very bright character'.[1] To look through Waterton's scrapbook of cuttings from his own articles is to marvel at his output. In many weeks he wrote almost daily, and much of the time he must have been at full stretch in order to get his stories and check his facts and sources. Two things in particular become clear. Foremost is that in those days the British public's appetite for any and all news about aircraft was truly voracious. And secondly, despite his reputation as an awkward customer who put bigwigs' backs up, Waterton evidently maintained good and cordial relationships with a network of friends in the RAF and people in the aviation industry as well as with fellow pilots. His log book shows that after he left Gloster he went on flying quite frequently on what look like grace-and-favour outings with old mates. It is hard to see how otherwise he could have been privy to the inside information that often enabled him to scoop his Fleet Street rivals.

His life had suddenly improved out of all recognition. Gone were the days of cheerless bachelor digs in

Cheltenham. He, Marjorie and their baby son Willy were now living rather elegantly in Cheyne Walk – for at last he had a job that paid a salary munificent by comparison with Gloster's pittance. He was doing something that interested him and about which he felt strongly, discovering along the way that he was not at all a bad journalist. He had a pilot's passion for factual accuracy that matched Christiansen's own and greatly pleased that editor, who hated making apologies for errors that found their way into print. He soon made friends with a fellow *Express* staffer, Tim Hewat. Hewat was an Australian journalist who at the age of only twenty-six was making a name for himself as the paper's maverick managing editor. Half a century later the author and film critic David Robinson was to describe Hewat in his obituary as 'colourful, outlandish, flamboyant, uncouth, unBritish'.[2] The last epithet alone would have given him common ground with Waterton, who was constantly aware of his own ambiguous nationality. Hewat was soon to move on to Granada Television where he devised the *World in Action* documentary series, but in 1954 he and Waterton must have agreed that the crashes of the hugely touted Comet airliner were the moment's obvious subject for someone of Hewat's investigative energy, with Waterton the obvious technical collaborator. The *Express*'s new air correspondent had, after all, flown the Comet himself. (His log book reveals that on 17 April he again flew Comet G-ALYT with John Cunningham and Cunningham's number two, Peter Buggé.) Accordingly, he and Hewat began collaborating on a book about the disaster that had so publicly overtaken de Havilland's premier aircraft.

The world's first jet airliner had made its maiden flight back in 1949. The de Havilland Comet could trace its origins to the wartime Brabazon Committee which had drawn up the specifications of some commercial aircraft which, it was hoped, might enable Britain to break into the United States' virtual monopoly in this type of aviation after the war was over. The Comet promised to make up for other projects that had so far conspicuously failed (such as the Brabazon) by leapfrogging even the best aircraft that the Americans had. This genuinely visionary aeroplane was designed to fly at speeds and altitudes unapproached by the propeller-driven Douglas DC-6Bs and Boeing Stratocruisers then dominating the trans-atlantic routes. Indeed, they were heights and speeds unknown in civil aviation and currently proving hazardous even to the early jet fighters. At the time it was by far the most rigorously tested aircraft ever to enter commercial service. In 1950 John Cunningham had taken the Comet on its first major proving flight overseas with eleven engineers aboard, non-stop from Hatfield to Cairo and then on via Khartoum to Nairobi. It was one of several occasions when Cunningham, a master of cool, took off, handed over to Peter Buggé and plotted their course using the tiny world maps at the back of his Letts pocket diary. His displays of the new airliner at Farnborough air show in 1950 and 1951 aroused huge public interest and enthusiasm.

As a result, much national tub-thumping heralded the Comet when it entered regular airline service on 2 May 1952. David Lean had already shot the aircraft as an icon of the future for *The Sound Barrier* (which premiered that same year). It appears in the film as a marvel of the age, not least because through a lapse in continuity G-ALYR magically morphs into G-ALZK within minutes at Cairo airport. But there was no gainsaying the sheer elegance of the Comet's clean lines, gleaming in mirror-finished alloy. Lean's camerawork as the aircraft taxies out emphasises the length and smoothness of the polished fuselage as it practically nuzzles the lens in what comes close to aviation pornography. The Comet was unlike anything else in the skies and seemed destined for enormous commercial success for at least the next six years, which was how long the abashed Americans themselves estimated it would take them to catch up. For the first two years of its service passengers were loud in the airliner's praise. It was so much faster, quieter and smoother than anything else they had ever flown in. The lumbering old Boeing Stratocruisers and Lockheed Super Constellations suddenly felt like cattle trucks by comparison.

Then it all went badly wrong. In October 1952 a Comet taking off from Ciampino airport in Italy failed to get airborne and crashed off the end of the runway. Two passengers were slightly injured but the aircraft itself was a write-off. Five months later a new Comet belonging to Canadian Pacific Airlines crashed in Karachi, also on take-off. This time all five Canadian Pacific crew and six de Havilland technicians died; the aircraft was burnt out. In May 1953, a year to the day after the Comet had inaugurated modern jet travel, one crashed in India in a violent tropical storm killing all forty-three aboard. In those days air crashes were comparatively frequent and it was accepted as one of those things although the accident was attributed, ominously, to airframe failure. But then in early 1954 the two Comets disappeared into the sea off Italy only three months apart and the possibility of a radical fault could no longer be ignored. Amid Air Ministry worries about Britain's increasingly questioned reputation as a military as well as a civil aircraft producer, the investigation was given to RAE Farnborough instead of to the Ministry's own Accidents Investigation Board. A Comet was exhaustively tested at Farnborough in a specially built water tank that subjected the fuselage to repeated cycles of pressurisation. Eventually a crack due to metal fatigue developed at the square corner of one of the windows: the source of the explosive depressurisation that had downed the two aircraft. The aircraft was redesigned with rounded windows and radically strengthened airframe and the Comet 4 resumed service with BOAC in 1958. But by then it was too late; the commercial lead was lost. In the interim the new Boeing 707 and Douglas DC-8 jets had appeared, both of them faster and cheaper to operate. It was a competition the Comet – no matter how improved – could no longer win. The American aircraft had wings with greater sweepback, giving them more speed. They could seat 180 passengers as opposed to the Comet 4's 100; and their podded engines were easily accessible, hence quicker and cheaper to service as well as being less of a fire hazard than the Comet's Avons, buried in the wing roots as they were. Boeing and Douglas executives and engineers readily admitted later that they had learned a great deal from the Comet's failure.

In the event, the Court of Inquiry held publicly into the accidents officially exonerated everybody. In private, though, it was known that people in the industry had from the beginning been worried about de Havilland's working in isolation on such a revolutionary aircraft. At the end of the war the Air Ministry had made all sorts of difficulties about 'demilitarising' the business of civil aircraft design and manufacture, refusing to allow its designers to work in the private sector. It was yet another example of the slowness with which Britain shed its war mentality long after the need for it was past. The result was that de Havilland took on an embattled mindset, even being described as 'uncooperative' by officials who wondered how the Comet's technical specifications could be independently checked. Sir Geoffrey de Havilland's famously autocratic rule probably did little to ease this impasse. Engineers at RAE Farnborough were reportedly concerned that the gauge of the aircraft's aluminium skin was too thin – worries well borne out by the remarks of de Havilland's test pilot John Wilson recorded in Chapter 1 (pp. 39–40).

Alas, no one persuaded Sir Geoffrey that he could have used a dispassionate second opinion. The American Federal Aviation Administration's own accident overview of the Comet 1 noted: 'Even large static strength margins do not negate the effects of fatigue . . . [T]he fatigue properties of the production airplanes were not well understood, and test results were misleading regarding the airplane's actual fatigue life . . . The test specimen had been previously cold-worked due to overloaded proof pressure testing. This overload inadvertently *enhanced* the fatigue life of the specimen, and thus produced results which did not reflect the production configuration.' In other words the various pressure tests de Havillands had repeatedly conducted on their samples of alloy had actually hardened them, giving the impression that untested production alloy was stronger than in fact it was.

In short, it looks as though the Comet disasters might have been avoidable had the aircraft been accorded full Air Ministry backing and co-ordination from the first, as befitted what was widely seen as the jewel of the nation's aero industry. Instead, lunatic demarcations of officialdom conspired with an inappropriate management system to prevent any easy consultation and input from engineers outside de Havilland. The resulting Comet was, as Bill Waterton was to write, 'a brilliant conception let down by its aerodynamics, engineering and handling – nothing like a hundred per cent aeroplane'.[3] He clearly knew a good deal that went unrecognised by the Court of Inquiry.

One of the things he knew was by personal observation. Many years later Waterton was to recall the flights he had made with John Cunningham in the early Comets. On several occasions he had noticed that especially during take-off the pilot seemed 'very busy and even uneasy' despite Cunningham's public contention that the Comet 1 and 1A flew 'extremely smoothly and responded to the controls in the best way de Havilland aircraft usually did'. But the two early crashes when the aircraft failed to take off revealed this as untrue. Take-off and landing were both far more critical than Cunningham, loyal de Havilland test pilot that he was, ever let on. The pilot in the Ciampino crash, R. E. H. Foote, was officially blamed for the accident and tried for years to clear his name. The pilot in Karachi was also blamed posthumously, in effect exonerating the Comet. Yet the aircraft was indeed at fault. If on take-off the pilot pulled the stick back ('rotated') even a fraction too far, the amount of wing area exposed in the angle of attack would cause the wing to stall and the airflow to the engines to be disturbed. If, on the other hand, he pulled back too little the aircraft would never lift off.

Such problems tended not to happen with propeller-driven aircraft, which were much more forgiving and had more 'feel' in their controls. It took several tragic losses of Comets for completely new instruments to be devised that gave the pilot accurate readings of his aircraft's critical pitch attitude. As one later test pilot sympathetically wrote:

Everything about the Comet was new and the players – industry, regulators, operators and aircrew – never fully realised that the old ways of doing things [i.e. as with propeller aircraft] were gone. It took a high body count to provide the impetus for major changes in flight testing standards and cockpit instruments. This is often the price of introducing new technology into aviation and is also one of the reasons why most people in the trade are very conservative and cautious about change. Also, the accident investigators [of the Comet crashes] had none of today's magic black box flight recorders to determine what had happened. It was all guesswork.[4]

The Comet had indeed been 'nothing like a hundred per cent aeroplane', but Waterton reserved this brutally truthful insider's verdict for the autobiography he was already compiling when his and Hewat's book, *The Comet Riddle* (Frederick Muller, 1955) was published. *The Comet*

Riddle was an even-handed popular summary of the aircraft's spectacular success and equally spectacular failure. It is still eminently readable. It stuck to the established facts and steered clear of making allegations. In this, no doubt, Tim Hewat acted as a restraining influence. The book sold well and was quickly reprinted. Since it was published three years before the Comet flew again the authors were not able to assess the disaster's longer-term implications. But Waterton would have sensed that the entire British aviation industry had suffered a blow, and it was one from which some feel it never completely recovered. The effects on morale were devastating, especially for de Havilland and his company. Whatever remarkable and original aircraft were still to appear – notably the English Electric Lightning and the Hawker/BAe Harrier – the heart had gone out of the buoyant, New Elizabethan optimism that had believed Britain was going to take charge of the world's air routes as in Tudor times she had its sea lanes. At the Court of Inquiry Lord Brabazon of Tara – Britain's first aviator and the man whose wartime committee had proposed the Comet's specifications – gave his bluff verdict in an address that must have been of dubious comfort to the victims' families. 'You know and I know the cause of this accident. It is due to the adventurous, pioneering spirit of our race. It has been like that in the past, it is like that in the present, and I hope it will be in the future.' Even today, the inflated rhetoric of business brochures ritually extols the merits of pioneering; but it is seldom mentioned that when pioneers fail it is usually to their rivals' direct advantage. So it was with de Havilland's lovely but tragic Comet.

For two years following his appointment Bill Waterton's columns mainly concerned themselves with aviation stories of the usual kind: the first flight of the Folland Midge (August 1954), for example; introducing the first pictures of the Fairey Delta 2 (October 1954); or announcing in the same month that plans for developing Gatwick as London's second airport had been given the go-ahead. But gradually, almost as if his confidence were growing or his patience wearing thin, there were more articles outspokenly critical of the British aviation industry, both military and civilian. His detractors would soon decry this as a personal vendetta; but the truth is that Waterton's criticisms were doing little more than keeping step with the increasing volume of alarm being voiced by MPs in the House of Commons, while spreading the word to the general public that all was not well with British aviation. At around this time, urged on by his wife Marjorie, he must have begun writing his own memoir, *The Quick and the Dead*. There are strong parallels between his journalism and many of the book's main themes. It sometimes seems as though he was using the news stories about how and why the country was falling behind its main ally (the US), its main enemy (the USSR), and its main competitor (France) in order to bring into focus what he had experienced for himself over many years. It was not just depressing but downright embarrassing to know that when the RAF flew its war-game exercises the pilots of the Canberras pretending to be incoming Soviet bombers had to be told to fly below their altitude and speed capabilities because otherwise the RAF's front-line fighters, Meteors and Vampires, couldn't reach them to launch their mock attacks. This was sorry stuff for an ex-RAF pilot to have to reveal, especially one who had himself flown real combat missions in Hurricanes and Spitfires only a dozen years earlier.

'Where Are the RAF Fighters?' was the belligerent headline to his column on 6 December 1954. 'The Hunter and the Swift have been produced in hundreds, yet not twenty are in squadron service. Meanwhile Fighter Command carries on with outdated and outpaced Meteors, Vampires and Venoms.' He notes that both the Hunter and the Swift have had frequent accidents, the Swifts being grounded for their controls to be modified and the early Hunters having been held up with under-carriage problems, to have air brakes fitted at RAF insistence, and because when they fired their Aden cannon the ingested gases caused their engines to surge. Even the old Venom had to be restricted in both performance and operation for safety's sake. 'Troubles are to be expected with new planes,' Waterton conceded, but 'Britain's new fighters are having more than their share.' Then the sting in the tail: 'Particularly as the latest RAF fighters are already outclassed by American types in service.' Two days later he returned to the attack. 'The Navy, as well as the RAF, is having trouble with new planes: four of its latest are not fit for battle.' He listed the current problems afflicting the de Havilland Sea Venom, the Westland Wyvern, the Fairey Gannet and the Hawker Sea Hawk.

'Finally,' he added once again, 'the Navy's fighters, like the RAF's, are outclassed by their latest American counterparts.'*

In due course Bomber Command, too, came into his roving sights. He noted that, as with fighters, there were serious delays in getting Britain's V-bombers into production. But by now it was clear that Waterton was no maverick, a lone voice crying doom from the sidelines. A White Paper on Britain's military aircraft, published in February 1955, was hotly debated in the House of Commons and Woodrow Wyatt MP called the £20 million spent on the Supermarine Swift (now effectively axed) 'an appalling waste of money'. Under pressure the Minister of Supply, Selwyn Lloyd, said, 'Our difficulties are essentially of development and not of production.' To which Waterton retorted in his column, 'I emphatically disagree with the Minister. Development and production are indivisible. Production in Britain is still by antiquated methods. Lack of modern tools and modern methods limits designs. I doubt if the tools exist in the British aircraft industry to build the American Sabre jet fighter.'5 As they *do* exist (he

---

* When Waterton complained – as he frequently did – of shoddy manufacturing practices in the aero industry, it should not be assumed that such problems were confined to Britain alone. In 1952 the American Grumman company had produced an advanced swing-wing experimental jet fighter, the F10F-1 Jaguar, that was plagued by sundry problems including some with its dreadful Westinghouse J40 engine that eventually led to the project's cancellation. After one near-fatal engine failure it was discovered that the access panel covering a box of electronics had been secured with three screws measuring less than half an inch and a fourth nearly five inches long that penetrated deep into the circuitry and was shorting it out. Amazingly, this proved not to be deliberate sabotage but simply slapdash work.

might have added) in Canada, where the best-finished Sabres of all were being built under licence to serve in the RAF until Hawker Hunters could replace them.

This was fighting talk, and he returned to the theme two months later, using a mix of rhetoric and unpalatable facts calculated to make the scrambled eggs fall unheeded from the mouths of empurpled gentlemen all over the Home Counties. The occasion was that of France's Sud-Ouest 9000 Trident, a mixed rocket-and-jet interceptor of advanced design, going supersonic even in a climb. 'Now the French Whip Us', was Waterton's column headline, adding that it was 'a black day for British aviation'. He put the story into context by first noting that by now, no fewer than *twenty-five* different American aircraft had flown supersonically.

The *only* British-designed plane capable of faster-than-sound speeds yet fit for squadron service is the Hunter. And it equips *only* four squadrons.

Yet, despite the manufacturers' fancy claims, supersonic flight with British fighters is still more of a stunt than a practical fighting manoeuvre.

Whose fault is it? Britain's engine manufacturers are not to blame. Rolls-Royce and Armstrong Siddeley engines power 2 French, 1 Swedish, 1 Australian and 4 American supersonic fighters.

So the blame must lie with Britain's plane-makers. They have stuck to out-of-date methods of construction that require thick wings for strength. But thick wings mean slower planes. And they have fallen behind in ways of controlling planes at super-speeds – so our planes do not compare in 'handling' with those of other countries.[6]

A month or so later he explained that the French secret of success was '*Ideas!*' At the end of the Second World War the French aircraft industry was left in ruins, but the French had been determined to restore it to its former glory. 'Dead wood was cut out. Young designers and engineers were encouraged... If foreigners had good ideas France took them, studied them, improved them – and used them in her own planes... *Yes, France has learned that IDEAS put a country ahead.*'[7] The comparison with Britain was all too shaming, he expanded. At the end of the war Britain already had the Gloster Meteor while the French had nothing. Within four years the French had flown the prototype of their Dassault Ouragan jet fighter that shortly afterwards went into service with both the French and Indian air forces. Exactly ten years later and having started from scratch, the French were now ahead of Britain. There was a further damning postscript when the US government's Hoover Commission questioned spending millions of dollars on British aircraft that didn't come up to US and Russian standards of performance and were anyway late in development. 'At the Paris Air Show in June,' Waterton added, 'I was told by an American Air Force chief: "The French put things right when we ask and are willing to make changes. The British won't do a thing unless they are forced to."'[8]

This was humiliating stuff for *Express* readers who (like me at the time) were still dreaming of British supremacy in the air as if the year were still 1946. Perhaps the real straw in the wind blew past almost unnoticed on 1 September 1955 with the news that the Society of British Aircraft Constructors had decided to ban supersonic bangs at that year's Farnborough air show. The SBAC's director, E. C. Bowyer, explained: 'No firm wants "bangs" this year. They no longer have any sales value.' It was a significant moment. Maybe what Waterton had dismissed as 'a stunt' was too loud a reminder of the effort still required for a British aircraft to perform it, while elsewhere the sound barrier had largely ceased to exist as an aerodynamic hindrance. Perhaps after all a great British illusion about its own aviation prowess had peaked and was beginning to wane. For some time there had been a rising chorus of complaint about sonic booms over England's overpopulated south and test flying was subject to new restrictions unthinkable in the glory days only four or five years earlier. This hindered the testing of genuinely supersonic new prototypes like the Fairey Delta 2 since by law they were not allowed to be flown over the sea in case they crashed and the flight recorders' data were lost. The ban on air-show bangs meant that we New Elizabethans were less likely ever again to thrill to the sight of an aircraft flashing low across an airfield towards us, partially cocooned in its own pale shimmer of condensation and almost silent until it had passed before battering our intestines with the sonic assault of its immediate climb. The unique sense that the vanishing dot was drawing something out of us and up with it into the sky would be missed, but never in a lifetime forgotten.

The Quick and the Dead was published in mid-1956. Formally it was Waterton's autobiography, but he can only have intended it as a *j'accuse*. Certainly that is how it was received. Not surprisingly, reviewers in general were a good deal less hysterical in their reactions than were the voices of outrage raised from the aero industry. The author had earned popular following and respect for his controversial journalism and most people seemed to recognise that in its often overstated fashion his book was founded on a deep and injured patriotism together with a voiced – as well as implied – loyalty to the RAF. This was a man who wanted the best for the Commonwealth's mother country and for his old service and had little but contempt for the bean-counters and bureaucrats who were nowadays allowed to stand in its way. Yes, Britain's defence industries were seriously short of funds, but political prevarication and the aviation industry's wilful refusal to reorganise itself were also wasting prodigious sums annually. To see how the book struck a close ally it is worth quoting *Time*'s review at length because it also neatly summarises Waterton's arguments.

In a new book *The Quick and the Dead*, a hard-hitting indictment of the whole industry, Bill Waterton charged that British aircraft firms, 'emasculated by safe government contracts', lack competitive drive. Fearful that the industry will be nationalised, they are less concerned with turning out fast airplanes than with turning a quick profit. As a result the industry is shackled by incompetent, underpaid employees, overlapping programs and antiquated factories that look like 'back-alley garages' beside US aircraft plants. Said Pilot Waterton: 'We have tried to muddle through by guess and by God. Britain [is] almost an also-ran in the aircraft stakes.'

The British, charged Waterton, are 'trailing behind America and Russia, which have both produced supersonic fighters in quantity and have bombers in service twice as big as our largest.'

Through lethargy and bad planning, Britain's planemakers have missed the rich civilian market for helicopters, light business aircraft and long-range jet airliners. Even if the British wished to introduce US designs, 'we haven't the means of transferring them to the production belt. We are building planes almost identically to the way we did 15 or 20 years ago.'

The main trouble is that 'few British firms understand development work.' Instead of trying to correct the deficiencies that show up in the prototypes, British aircraft 'boffins,' i.e. chairborne scientists, try to cover up to save costly redesigning. Despite the industry's often brilliant performance at Britain's annual Farnborough air show, Waterton points out that the show is 'a lot of sham.' The aircraft entered are often prototypes, years from the production line and often perilously under-tested. Says he: 'It is a miracle that there are not mass disasters at Farnborough every year.'

As a case history of boffin botchery, Waterton cites the Gloster Javelin on which he did all the initial testing. Despite a crash in which he almost lost his life, he said that his criticisms of the plane were generally ignored until Gloster's No. 2 test pilot was killed in a Javelin crash. Gloster replied to Pilot Waterton's blast last week with the countercharge that he had not quit but was fired for his 'disinclination to continue the necessary research flight-testing of the Javelin,' and dismissed his book as a mishmash of 'harrowing self-dramatisation, sensational slanders, half-truths, recriminations and flaunted betrayals.' But Waterton refused to back down. Said he: 'I say, appoint an impartial commission to go into the whole matter and look at the records.'[9]

The Minister of Supply, Reginald Maudling, refused to set up an official government inquiry into the book's allegations – probably wisely, given that the evidence could only have confirmed them and added to the sense that Britain's aviation industry and air defences were in deep disarray. As regards his mention of the Javelin, *Time*'s writer most probably checked his facts with the two visiting American pilots who had flown the aircraft while Waterton was still at Gloster. If so, he would have discovered that their opinion tallied exactly with Waterton's: that it was a potentially good aircraft still very far from being ready for production. And as for its makers' bluster, Wing Commander Derek Collier-Webb's 1998 article 'Tested and Failed: Gloster Javelin' makes it clear that two years after Waterton's dismissal, men from the Ministry of Supply went down to Brockworth and read the company the riot act, issuing an ultimatum that it was 'a question of now or never': either Gloster got the Javelin right immediately or it would follow the Swift into oblivion.[10] They didn't get it right immediately but they did get it into production.

As the ancient Greeks noticed, the bearer of bad news usually gets it in the neck. Waterton's book finished off his own career as a journalist. A reluctant Christiansen was obliged to sack him because various heavyweight advertisers threatened to withdraw their custom from the paper unless he went. 'I got a tremendous amount of flak,' Waterton recalled in 2003. 'I knew hell would be raised. But I thought I would get some support from the Royal Air Force. Which I didn't. I thought somebody would say, "Here's a bloke that's trying to do us some bloody good."'[11] His departure merited a front-page headline: 'We Sacked Waterton – and Why'. That was in July 1956. Nobody can have been consoled when a bare three months later the *Daily Mail* printed extracts from the current issue of the authoritative American journal *Aviation Week* that completely bore out the substance of Waterton's allegations in the rival *Express*. From the freezing depths of the Cold War the journal's editor, Robert Hotz, made three sobering points:

1. The RAF is, and is likely to be for the foreseeable future, saddled with obsolete aircraft.
2. The present rate of progress in the French aircraft industry may push the RAF into fifth place in terms of the quality of its equipment.
3. Britain's guided-missile situation is 'even worse' than its fighter position. It is 'lagging at least two development generations behind the US'.

Hotz added: 'We do not recite these sad facts to twit our British friends. It is not a pleasant sight to watch an ally whose military air power record is so distinguished slide down the technical slope towards obsolescence.'[12]

I first came across *The Quick and the Dead* either that year or the next in the Canterbury branch of Boots' Booklover's Library, took it away and devoured it. Some of my precious illusions about the state of Britain's air defences had already been badly shaken but I still found Waterton's story devastating. At the same time it greatly appealed to my adolescent pleasure at seeing authority and the powers-that-be so thoroughly mocked (one pleasure that has lasted a lifetime, it turns out). My juvenile hero worship turned into a genuine esteem for Bill Waterton which, unlike most other youthful enthusiasms, has never really waned because subsequent history proved he had been right all along. It was only sad that his accusations came too late for Britain's aviation industry to pull itself together and inject some French-style energy and efficiency into its proceedings.

After ten years in the public eye Bill Waterton abruptly disappeared from view. It is hard not to see him as the sacrificial scapegoat, heaped with the aero industry's sins and embarrassments and driven forth into the desert with the Gloster Aircraft Company's imprecations ringing in his ears. Once the *Daily Express* had sacked him in that summer of 1956 he and his family left Britain with a typical parting shot. 'I don't care what they say about me – I'm not important,' as Gloucestershire's leading newspaper quoted him. 'What does matter is that red herrings should not be drawn across the trail. If I have drawn attention to unhappy things going on in the aircraft industry and can help to get them changed, I shall be happy.'[13]

Then he vanished; and for the next fifty years utter silence surrounded the name of my boyhood hero.

Bill Waterton's glum perception that there were political, industrial and managerial matters seriously wrong at the heart of British aviation in the mid-fifties was unfortunately all too well borne out by subsequent events. Even so, it is possible to argue that his personal viewpoint was particularly jaundiced by his experiences at Gloster. The Meteor's early success does seem to have made that company self-satisfied and indolent in a way that Waterton obviously came to feel was peculiarly British, in that having taken the lead they sat back rather than redoubling their efforts to develop and improve their aircraft while moving swiftly on to the next project. Yet not all British aero design departments stubbornly ignored what their chief test pilot was saying even when the company was a fellow member of the Hawker Siddeley Group. At Avro, for instance, the test pilot Roly Falk (he of the pin-striped suits) insisted on designing much of the Vulcan's cockpit layout himself and got his way; whereas Gloster's design department seems consistently to have refused to listen to Waterton on the same subject. That company's failure to produce another aircraft with their name on it after the Javelin is surely significant.

Other test pilots at other companies had a rather less gloomy take on the period. In Roland Beamont's case, for example, this was perhaps due to a sunnier temperament and even to the astounding luck of test flying three of the most remarkable British aircraft of the period: the Canberra, the Lightning *and* the TSR.2. However, if his and Waterton's experiences look as though they represent two extremes of a test pilot's fortunes in the British aero industry of the period, somewhere between them lies an irreducible melancholy truth. Although throughout the industry designers of genius were allied with inspired engineers to produce brilliant ideas, full-scale mock-ups and even prototypes of aircraft with considerable potential, practically all these projects were destined to be fatally delayed by official indifference or indecision; else they were starved of funds, aborted by abruptly changed specifications or just wilfully cancelled at the last minute. The dreadful saga of the TSR.2 will be recounted shortly; but before that, the story of the Fairey F.D.2 poignantly illustrates the 'If only . . .' quality that mournfully adheres to so many projects of the period.

If ever an aircraft looked aesthetically right the Fairey Delta 2 did. It was an elegant, slender aircraft with neat wing-root intakes, a steeply raked squared-off tail fin and a long needle nose. Since this nose was practically all the pilot would be able to see when landing or on the ground, the F.D.2 was given a 'droop snoot': a pioneering piece of British design later incorporated into Concorde. The entire nose was hinged behind the cockpit and could be hydraulically tilted downwards by ten degrees to give the pilot full view during taxiing, take-off and landing. And as if in answer to Waterton's insistence that the thick-winged Javelin would never be a proper supersonic aircraft the F.D.2's designers in the late forties made it the first British thin-winged delta. The contract for the experimental prototype was signed in 1950, but Fairey was a comparatively small company with limited manufacturing capacity and priority had to be accorded to its anti-submarine turboprop Gannet, which the navy were in urgent need of. Thus construction of the F.D.2 began only in 1952. What had started as an extremely advanced design was already delayed by two years.

The aircraft was flown from the beginning by Peter Twiss, an ex-Fleet Air Arm pilot of great talent, who soon found it handled as well as it looked. That was until the fourteenth flight in November 1954. Shortly before take-off from Boscombe Down it had been decided to remove a small rubber sealing strip from between one intake and the fuselage because Twiss thought it looked like coming loose and he was concerned that it might be sucked into the engine. He got airborne and climbed to 30,000 ft, going like a bird, when he happened to glance at the fuel gauge and saw it unwind in a matter of seconds from 'Full' to 'Empty'. Since he had been in the air only a matter of minutes he naturally assumed the gauge was faulty, but being a professional he took the precaution of calling Boscombe to get a home bearing in case of emergency, never seriously believing he would need it. Moments later the fuel pressure warning light came on and the engine spooled down and stopped. 'This left me at 30,000 ft, still flying away from Boscombe Down, but with no engine. The Delta was now a glider, and a pretty fast one at that.' Twiss turned back to the home bearing he had been given and faced the same dilemma as Waterton had two years earlier in the Javelin: could he get this valuable prototype back in one piece or would it be safer to eject? Waterton had still had his engines and was able to gain at least minimal control even without elevators. Twiss's problem was that the F.D.2's controls were hydraulically powered and could not be operated manually in the event of hydraulic failure. Without the engine running there was no way of generating hydraulic pressure. There was still some pressure in the system, but once those reserves were used up he would have no means of controlling the aircraft, by which time he could be too low to eject safely. Twiss calculated that he might just have enough hydraulic reserve left to get him back to Boscombe, but it would be a gamble.

At about 2,500 ft the F.D.2 suddenly broke cloud and he was reassured to see the runway nicely lined up six or seven miles ahead. His new dilemma was whether to lower the droop snoot; doing so would enable him to see what he was doing when landing, but might also use up the last of his hydraulic pressure. He would be coming in far faster than usual since he dared not deploy the air brakes, which were also hydraulically operated. Delta aircraft flare for touchdown at a high angle, which was exactly why the droop snoot had been invented. But now Twiss decided he dared not risk lowering the nose. It would be bad enough to lose vision, but losing control entirely would definitely be fatal. He managed to get the nose wheel down but there was not enough pressure left to lower the main under-carriage. As he crossed the numbers at the end of the runway he found himself lying on his back and staring helplessly up at the sky as he touched down at 240 mph. He heard the tail hit first and then the Delta fell on to the nose wheel and tore down the runway with a grating scream, trailing sparks like a meteorite. It seemed to Twiss as though the speed would never lessen, then the aircraft suddenly veered off on to the grass and ploughed along with a great jolting that made him fear it might trigger the ejector seat. At about 50 mph he pulled the lever which jettisoned the canopy. 'I was out of that cockpit almost before the aircraft had stopped moving. From the time that the engine had stopped at 30,000 ft to the time I had jumped from the cockpit was no more than six minutes. But even looking back on it now, it seemed like an eternity.'

When he was satisfied that the aircraft wasn't going to explode, Twiss returned to the cockpit and made the ejector seat safe. He then checked the fuel tanks in the wing. They were full. The engine had been starved of fuel.

The cause was traced back to the gasket that had been removed before the flight. At high speed air had been forced into the fuselage, building up enough pressure to collapse a rubber tank and jam the fuel valves, shutting off the flow and interfering with the gauges. As Twiss later said, 'Who could have foreseen that the harmless, unwanted piece of rubber seal we had removed could have caused so much trouble?' The aircraft was badly damaged in the landing, but not irreparably so. Still, it was another eleven months before Twiss could take the refurbished F. D.2 effortlessly past Mach 1 in level flight. 'If ever the Delta had had to justify herself to me after her crash, this wonderful little aircraft did so that morning when she flew as gently as a bird into the hard supersonic October sky. From that moment, I knew we had a world-beater.'[14] Like Waterton, Twiss was officially honoured for bringing back his aircraft in one piece, receiving the Queen's Commendation for Valuable Service in the Air. It had all been done in less time that it took Britons on the ground to brew a pot of tea, or queue at the fishmonger's for some whale meat for the dog and coley for the cat, or nip to the newsagent's on the corner for the evening paper – and none of them any the wiser as to the drama that had been going on in the sky overhead.

It was soon obvious to Twiss that in the F.D.2 Fairey had a real contender for the world air-speed record, which in late 1955 was the American Colonel Horace 'Dude' Hanes's 822.26 mph in the Super Sabre: a speed that had seemed to us schoolboys slightly unreal, being nearly 100 mph faster than Neville Duke's previous record of 727.63 mph. Twiss was certain that the F.D.2 could bump the record up above the magic 1,000 mph. Yet only nine years after huge official weight had been thrown behind the High Speed Flight of Donaldson, Waterton and Duke with claims that national honour was at stake, there was now complete indifference at governmental level towards breaking the record. 'By far the hardest part of the whole project was persuading the Ministry that this was a thing we ought to do,' Twiss wrote later.[15] Yet he managed it; and in the following March he set a new speed record at 1,132 mph. The British press, by now used to expecting nothing but bad news from the aircraft industry, was largely ecstatic. Because Fairey had gone to great lengths to keep the attempt secret, mostly for fear the Americans would immediately make an attempt of their own, the whole thing came as a complete surprise. In a gentlemanly gesture, 'Dude' Hanes cabled his congratulations.

Now convinced that the F.D.2's speed could be still further increased and that the aircraft had great potential, Fairey and Twiss agreed that the next step would be to develop this prototype into a first-rate fighter. Although the Minister of Supply, Reginald Maudling, had recently hinted that the future defence of the realm might rely chiefly on missiles, various high-ranking RAF officers convinced Fairey to go ahead with the project without further delay on the grounds that a government could always be more easily swayed by an aircraft that had already been tested and was ready to go into production. What followed could scarcely have been more discouraging for the manufacturer of a potential world-class fighter. Because of the new regulation forbidding supersonic flights over 'populated areas' of the UK at altitudes of less than 30,000 ft, Twiss was driven abroad to do much of the F.D.2's testing in France. (Marcel Dassault, eponymous head of the best-known French aircraft company of the day, made him and his team very welcome at Cazaux, south of Bordeaux.) Next, the British government refused to pay for any damage caused by the aircraft's sonic bangs. When private UK insurance companies quoted exorbitant premiums that Fairey couldn't possibly pay, Dassault suggested a French insurance company that covered all the tests for £40. As Twiss summed it up: 'If it seems to you a pretty irresponsible way for one of the world's leading aeronautical nations to carry out the kind of development tests so vitally necessary if it is to continue being one of the world's aeronautical nations, I can only agree with you.'[16]

When Harold Macmillan took over the premiership following Anthony Eden's post-Suez disgrace and break-down, he appointed Duncan Sandys (pronounced 'Sands') as his defence minister. Sandys's main remit was to stem the haemorrhage of defence costs, above all in aviation, since it was clear that the long-overdue rationalisation of Britain's aircraft industry was now an absolute priority. As a major in the Royal Artillery during the war Sandys had commanded Z Battery at the missile testing grounds at Aberporth near Cardigan, an establishment closely connected with RAE Farnborough. He became increasingly sure that the future of artillery lay more in rockets than in guns, and his later experiences in London of Hitler's V-2 rockets convinced him of it. On 4 April 1957 he published a White Paper which in its way did more to inflict radical damage on the British aircraft industry than anything Hermann Goering's Luftwaffe had done over nearly six years of war. The section on research and development stated matters unequivocally:

In view of the good progress made towards the replacement of the manned aircraft of Fighter Command with a ground-to-air guided missile system, the RAF is unlikely to have a requirement for fighter aircraft of types more advanced than the supersonic P.1 and work on such projects will stop.

The P.1A was the Lightning prototype, English Electric's answer to the same Official Requirement that Fairey had met with their Delta. Elsewhere, SAMs (surface-to-air missiles) would take over from fighters, and Thor and Blue Streak missiles from bombers, so that presumably the days even of Britain's V-bomber force were already numbered just as the Vulcan and the Victor were entering squadron service.

At the White Paper's ground zero the immediate damage was enormous, irrespective of the longer-term fallout. Only English Electric emerged unscathed with a grudging reprieve. Roland Beamont's characteristic luck had held yet again, and by chance he took the Lightning prototype on its maiden flight on the very day the White Paper was published. Together with the same company's TSR.2, it was the sole aircraft that Sandys exonerated because 'unfortunately it has gone too far to cancel'. One might have thought the same exemption would have applied to Fairey's F.D.2, but Sandys did not. Despite holding the world air-speed record it had no future in the new dispensation. Equally convinced of their own product's sunny prospects were Hawker Siddeley, who had risked £1 million of the company's money (a huge sum in the fifties) on developing their P.1121, a potent supersonic interceptor. It, too, was cancelled forthwith. So was Saunders-Roe's SR.177, which was a mixed jet and rocket-assisted fighter in which both the RAF and the Admiralty were keenly interested. As for Avro, they lost two potentially major aircraft, one being the 720, a delta interceptor somewhat like Fairey's. The prototype was complete even to its RAF roundels and serial number, but the Ministry refused all requests to fly it and ordered it destroyed. The product of years of teamwork and skill, it probably went to the gunnery ranges at Shoeburyness.

The fallout from Duncan Sandys's White Paper was no less devastating than its immediate destructive power. One has only to read the books by the pilots, designers and company men of the time to appreciate the anguish and trauma it caused. Virtually everyone affected recognised it as a gross strategic blunder that would soon have to be reversed, but meanwhile it caused irreparable damage. Not only did it ensure that yet another generation of Britain's fighter aircraft would be forfeit, but in the enforced break-up of close-knit teams something was lost that could never be replaced. Whatever shreds of confidence in politicians the industry still retained vanished within hours of the White Paper's publication. It made clear in black and white Stationery Office print that Britain no longer wanted manned warplanes, and by extension it no longer wanted the industry that built them. Within a mere dozen years of the war's ending the RAF was effectively to be demoted from being the gallant last-ditch defender of the realm to a state of redundancy or – at best – a bunch of ground crews manning missile bases.

Such, at least, were the opinions voiced in the newspapers. But the fallout didn't stop there. The White Paper was the cack-handed weapon that finally forced a long-overdue series of mergers and takeovers on the aero industry by making clear that in future the government would only sign contracts with large groups of manufacturers and not with individual companies. It thus enforced the step that should have been taken back in 1945 when everything was in flux, of reducing the size of the industry while rationalising the way it was organised. In the chaotic hiatus that followed the war this could probably have been made acceptable had it been done in a firm and purposeful manner with a clearly understood objective. But Duncan Sandys's brusque ultimatum at a moment when companies were already demoralised by Whitehall's lack of direction was to have fateful consequences for the way the surviving industry functioned.

Amalgamations now became official policy. In 1959 Armstrong Siddeley and Bristol's engine division merged to become Bristol Siddeley – which was itself destined to be swallowed up by Rolls-Royce in 1966 to leave Rolls-Royce as the sole surviving major UK aero-engine manufacturer, which it remains to this day. In 1960 virtually all Britain's venerable aircraft companies were summarily bundled up into either the British Aircraft Corporation or Hawker Siddeley Aviation. BAC was composed of English Electric, Vickers-Armstrong, Percival, Bristol Aircraft and Supermarine. Hunting joined later that same year. Meanwhile de Havilland,

Blackburn and Folland merged with the existing Hawker Siddeley Group (which as we know already contained Gloster, Avro, Hawker and Armstrong Whitworth) to form HSA, Hawker Siddeley Aviation. The only major independent airframe company left outside for a while was Handley Page. (BAC and HSA were to be nationalised in 1977 as British Aerospace, BAe, which continues today as BAE Systems: the single remnant of the country's once motley collection of private airframe companies.) As for the wretched Fairey, they were sold off to Westland, who were really only interested in helicopters. Westland had their eye on Fairey's Rotodyne: an advanced autogyro-style transport helicopter intended for feeding airline passengers directly from airports to city centres. They didn't want the beautiful F.D.2.

To be fair, such awful strategic errors were not confined to Britain; Sandys was not alone in predicting the missile as the end of manned aircraft. In the United States there was a similar current of opinion that dogged the development of North American's advanced B-70 bomber, the Valkyrie, which suffered a very British-style history of stop-go indecision and policy changes. But American post-war aviation was always going to be able to withstand such setbacks without sustaining serious damage because it was backed by nearly limitless defence contracts, not to mention patriotic pride and pork-barrel funding from a government keen to play on it. Nor was there ever any hesitation in America's steady development of jet fighters to match those being produced in the USSR. Aircraft companies in the US never had to deal with anything remotely on a par with Westminster's graceless cheese-paring that obliged family companies like Fairey to insure at their own expense a top-secret experimental aircraft that the Ministry of Supply legally owned, and then take it to a foreign country to test.

In thinking that manned aircraft could be made obsolete by technology, if not by strategic need, Duncan Sandys may eventually be proved right, as today's UAVs (Unmanned Aerial Vehicles) are beginning to suggest.[*] As this is written, reconnaissance or combat drones in the skies above the Middle East are being remotely 'piloted' by air force personnel sitting at computers seven thousand or more miles away in the United States. But such a degree of development in electronics and avionics was as fanciful in 1957 as the anti-gravity boat on which the evil Mekon rode through Dan Dare's adventures in the *Eagle*. Sandys hardly deserves credit for foreseeing 'robot planes', which had long since been the stuff of science fiction. Over half a century ago his mission was quite simply to slash costs by substituting a cheaper alternative to manned aircraft – an alternative that, even had it worked as planned, was quite incapable of fulfilling either Britain's NATO obligations or the varied functions of combat aircraft during the nuclear Cold War. The whole point about Mutual Assured Destruction was to maintain a careful stand-off. A small part of this involved sending RAF fighters up on a daily basis over the North Sea to meet the Soviet Bear bombers that skirted British airspace, the aircrews often waving cheerily to each other as if in recognition of their respective

---

[*] In July 2009 the Minister for Defence Equipment and Support, Quentin Davies, said at an Unmanned Air Systems exhibition at the MoD that 'beyond [the 2030s] the name of the game will be UAVs'.

governments' lunacy. Sandys never explained how such encounters could be so amiably and safely achieved by missiles.

The final irony of the F.D.2 saga was the prodigious international success of Dassault's Mirage III fighter – which turned out to bear a considerable resemblance to Fairey's aircraft. As Derek Wood noted,

The success of the F.D.2 [in gaining the world speed record] confirmed Dassault's theories and [its] arrival at Cazaux provided very useful information which was fed into the Mirage programme ... The enormous potential of the F.D.2 concept was never used for a production aircraft and the harvest was left to France to gather. It is recorded in *Mirage, Warplane of the World* by Jack Gee that 'Later, when the Mirage was conquering markets all over the world, Dassault told a British aircraft chief, "If it were not for the clumsy way in which you tackle things in Britain, you could have made the Mirage yourselves."'[17]

At the time there were dark murmurings suggesting that Dassault's willingness to grant the British aircraft test facilities at Cazaux (where there was also a major French air force base) had been *perfide France*'s plot to steal the F.D.2's design. This was nonsense, of course. As Bill Waterton had rather tactlessly pointed out in his column, France by then hardly needed to crib from British aeronautical know-how. Marcel Dassault was as brilliant as any British designer, as his best-selling Mystère and Mirage I fighters had already showed. But there is no doubt that the F.D.2 helped him with details, as he freely admitted, and nor is there much doubt that if the British government had given the sort of patriotic backing the

French aero industry enjoyed, Fairey's machine could have achieved comparable success.

If only . . . Peter Twiss never again flew the F.D.2. After the White Paper a sole prototype eventually went to BAC as the BAC.221 and was used for testing Concorde's ogival wing shape as well as the design of its droop snoot. It was just as well that Bill Waterton had left the country by the time Duncan Sandys's axe fell. His column would surely have been written in sulphuric acid and – to make it still worse – with absolutely no sense of surprise. I like to think some of the White Paper's battered victims took pleasure half a dozen years later in the awesome scandal in 1963 of the divorce of the Duke and Duchess of Argyll, much of whose evidence centred around some lurid Polaroid photographs found in the Duchess's boudoir, one of which was rumoured to be of the man *Private Eye* promptly immortalised as 'Sunken Glandys'. This identification wasn't confirmed beyond reasonable doubt until a TV documentary in 2000, but there was no lack of people to believe it back in 1963. I suspect not many people in the aviation industry at the time read *Private Eye*, which is a pity. And I can't help adding the old saw about it being an ill wind, because one of the Minister of Defence's cost-cutting decisions in his 1957 White Paper was to do away with National Service, which I thereby missed by a whisker. Whatever harm he did to Britain's aircraft industry, old Glandys did me a bit of good by saving me from being banished to Caterham or Catterick, profitlessly peeling potatoes and bayoneting sandbags for two years, and I should be churlish to deny it.

'A power of no good'

In 1948 Nevil Shute published his novel *No Highway*. Appropriately for a novelist who was also a trained aircraft engineer, it is a thriller whose story concerns the fortunes of a new transatlantic airliner, the Rutland Reindeer, which turns out to have a metal fatigue problem. A boffin at RAE Farnborough has an entire tail section of the aircraft, covered with strain gauges, up on a test rig that vibrates it around the clock to simulate normal cruising flight. According to his newest calculations the front spar of the tailplane could fail without warning after 1,440 hours' flying. A Reindeer has already disappeared over Labrador with forty passengers but since the wreck is inaccessible it has never been investigated to establish the cause of the crash. It is now beginning to seem highly significant that the aircraft already had 1,393 hours' flying time. Should the entire Reindeer fleet be pre-emptively grounded, with dire consequences for the Rutland Aircraft Company and the airline, not to mention the risk of Britain losing its slice of the transatlantic route? An additional muddying factor is that it looks as though the boffin, Mr Honey, is also interested in pyramidology. But need this apparent evidence of nuttiness in private life invalidate his professional conclusions as a structural engineer?

Assuming *No Highway* was written the year before it was published, the story pre-dated the Comet 1's maiden flight by two years, and Shute's foreshadowing of the Comet accidents of 1954 seems almost uncanny. As a thriller it reveals the extent to which the safety issues of long-distance passenger flights were both popularly understood and debated. Its publication year also happened to mark the mysterious disappearance of the British South American Airways' Avro Tudor *Star Tiger* somewhere near Bermuda with thirty-one passengers and crew. The Tudor was a typical British post-war stopgap airliner: a derivation of the Lincoln bomber with the same wings but a new pressurised fuselage. As a passenger aircraft it was dreadful, right down to the very name with its absurd aura of half-timbering and National Trust. Even its designer, Roy Chadwick, who had shaped the brilliant Lancaster and Lincoln bombers, was at a loss to explain quite why the Tudor was so bad. The final irony is that it killed him. In August 1947 he took off from Avro's Woodford airfield in a Tudor with Avro's chief test pilot Bill Thorn at the controls and the aircraft abruptly nosed over and plunged into the ground, killing them both. It turned out that this was not due to a design fault but to an elementary maintenance mistake caused by a careless mechanic who had crossed over two control cables (which in turn suggests that Thorn might have skimped his pre-flight checklist). This did nothing to shift a feeling that the Tudor was jinxed as a passenger aircraft, even if it went on to give good service flying cargoes in the Berlin airlift.

In the BBC Radio 4 series *Inside the Bermuda Triangle*, broadcast in 2009, former BSAA pilots Peter Duffey, Don

Mackintosh and Gordon Store were interviewed and testified to design faults in the Tudor so elementary and so gross that it seems a misnomer to call them 'faults' at all – a word that can imply unfortunate oversight as opposed to reckless indifference. Gordon Store admitted that he had never had any confidence in the arrangement of the Tudor's Merlin engines. 'Its systems were hopeless . . . All the hydraulics, the air-conditioning equipment and the recycling fans were crammed together underneath the floor without any thought. There were fuel-burning heaters that would never work.' That heating system, as Don Mackintosh said, 'bled aviation fuel [i.e. 100-octane spirit] on to a hot tube – and was also fairly close to the hydraulic pipes'. It is now thought likely that *Star Tiger* and its sister *Star Ariel* (which vanished in the same area a year later) might either have run out of fuel well short of Bermuda or else have gone down in flames. It was these disappearances that helped give rise to the 'Bermuda Triangle' myth. One final accident sealed the type's fate when in 1950 a Tudor V, *Star Girl*, crashed on landing at Llandow in South Wales. A veteran of the Berlin airlift, it was returning, packed with triumphant rugby supporters, from Belfast where the Welsh team had just won their eighth Triple Crown. Out of seventy-eight passengers and five crew only three supporters survived. The loss of eighty lives was the worst civil air disaster to date.*

* It is strange and unjust that this appalling accident is virtually forgotten today, while the Munich air crash of 1958 with its far fewer fatalities (seven Manchester United footballers and fourteen others) is endlessly referred to, memorialised almost with relish, and may yet become the subject of a Hollywood film. Possibly this merely reflects Rugby Union's inferior popular status.

The point here is the contrast between the fictional Reindeer and the factual Tudor. Shute writes in his Author's Note: 'The scrupulous and painstaking investigation of accidents is the key to all safety in the air, and demands the services of men of the very highest quality. If my story underlines this point, it will have served a useful purpose.' It would be too easy to excuse the Tudor's shortcomings by saying, well, the aircraft had been designed a year before the war ended; in those days London–Bermuda with a refuelling stop in the Azores was right at the limits of almost any commercial aircraft's range; and standards were anyway a bit rougher-and-readier then. Yet once BOAC had swallowed up BSAA its extreme reluctance to accept that airline's Tudor fleet – understandably preferring tried-and-tested American DC-4s and suchlike – made it clear that even in those days passenger aircraft were expected to be reliable and safe as well as economical, just as Shute's novel implied. Moreover, it was widely understood that any failures would be rigorously investigated by real-life Farnborough boffins. The public in 1948 may have been more phlegmatic about flying in general, especially when faced with extreme weather conditions, but that didn't mean passengers easily accepted they might die if they flew. Any airline hoping to capture a share of the profitable transatlantic market understood very well the trade-off between spending for greater safety and the penalties for gambling with people's lives.

Even discounting safety issues, Britain's approach to gaining a foothold in post-war commercial aviation was strangely muddled and half-cocked from the start. The

United States held a near-monopoly on modern long-distance passenger aircraft, partly from having to deal with its own vast continent but also because it had been able to go on developing them throughout the war. That Brabazon and his committee – not to mention politicians like Churchill – should have taken it for granted that Britain ought to challenge this superiority indicated hubris as much as it did serious consideration of the economics involved and the chances of real money-making success. By 1943 one hardly needed to be psychic to predict that once hostilities had ceased Britain, with its war debts to pay off, its huge overseas empire to maintain and its own partly ruined cities to rebuild, was going to be dangerously short of money. It might have occurred to these patriotic gentlemen that it would be judicious to abandon to their vastly wealthier ally a field of aviation in which they already lagged far behind, while they concentrated on maintaining their own genuine lead in jet-fighter technology. The task of building competitive commercial aircraft was formidable. Nor was it just a matter of retooling Britain's war machine to build civil aircraft: the industry would have to change its entire attitude, especially where safety was concerned.

The Comet's failure to live up to its revolutionary promise, while lamentable, was not completely surprising. Despite de Havilland's care and attention to detail – and in terms of progressive engineering the company at the time was probably the closest of all British aero firms in spirit and practice to an American counterpart – the Comet still went into service when it was 'nothing like a hundred per cent aeroplane'. Although as we have seen this was not an uncommon practice with British military aircraft, and the same companies were building both military and commercial models, it is still to be wondered at. Even at airline level the instinctive reaction to disaster was to close ranks. Thirty-seven hours after the first crash Sir Miles Thomas, BOAC's chairman, temporarily suspended all Comet flights 'as a measure of prudence' to enable 'minute and unhurried technical examination of every aircraft in the Comet fleet to be carried out at London Airport'. But then BOAC rather spoilt the effect by announcing that its chairman's decision was 'based on a desire to retain the good name of the Comet'.[1] One might have expected him to have valued his passengers' lives over an aircraft's reputation, at least in public.

The attitude of some companies towards the proper way to build and test an airliner intended for mass airline sales was wrong. After all, with the exception of possible export sales military aircraft were not designed primarily to be an economic success. So long as their performance specifications were met and the government's cheques rolled in on time, the companies responsible for them had little incentive to consider their aircraft from an economic viewpoint. Issues such as fuel consumption, let alone noise levels, never came into the equation. Nor was safety always of the paramount importance the companies ritually claimed. As we know, extraordinary risks were taken with barely airworthy prototypes such as the Javelin being rushed into the air to meet some financial deadline, while operational RAF squadrons kept their stocks of black sealing wax constantly replenished.

Such habits were precisely the wrong ones for any company to bring to its commercial designs. So also were production practices that Bill Waterton had dismissed as being more typical of 'back-alley garages'. His was by no means a lone voice, especially among people who had travelled or worked abroad and had seen how differently things were done elsewhere. By 1953 the Canadian-born Beverley Shenstone, the aerodynamicist who had helped design the Spitfire's elliptical wings in the mid-thirties, was British European Airways' chief engineer. At a conference that year he was outspoken:

In the United Kingdom, the average finish given to an aircraft is far inferior to that given to the average American aircraft. Parts and assemblies are not treated well in the shops and in one wing factory the impression was that after manufacture they dragged the wings along the shop floor.[2]

It was to be hoped that this sort of lackadaisical crudeness was the exception rather than the rule. But the real blame for what blighted several potentially great British commercial aircraft lay far more at the level of management. After all, they had the power to change shop-floor practices overnight had they really wanted to.

What also seems incomprehensible today is the creaking *slowness* that attended virtually every step of Britain's aircraft production and anything else to do with aviation. Consider, for instance, the post-war development of Heathrow. In 1946 what had been Fairey's Great Western Aerodrome was demilitarised and reborn as the fledgling London Airport. From the beginning it was billed as 'the world's largest airport and the country's biggest post-war building scheme', and was clearly intended as the flagship civil facility for a rapidly growing peacetime aviation industry. Within seven years it was handling a million passengers annually, even though these unfortunates still had to troop in all weathers through a series of tents instead of a terminal. As befitted a true-blue New Elizabethan plane-spotter, my bookshelf contained a copy of *London Airport: The official story of the new world air centre* (HMSO, 1956). The pretext for this glossy booklet, halfway between a photo-essay and a guide, was the opening of the Queen's Building by Her Majesty in December 1955. It illustrates the airport's new amenities (project architect: Sir Frederick Gibberd) as well as constituting not very subtle propaganda for the British aircraft industry. However, amid all the flag-waving for this supposedly state-of-the-art 'world air centre', there is conspicuously not a single mention anywhere of a rail link even as a future project. Passengers either arrive 'by airline coach from the Air Terminal in London' or else 'independently' – meaning also by the pre-motorway, traffic-clogged 'Great South-West Road'.*

Many of *London Airport*'s pages carry photographs of aircraft taking off or landing, or else merely sketched in a maintenance hangar. Otherwise they are shown standing outside the new terminal building (not many passengers

---

* It seems inconceivable that it should have taken over thirty years to build any sort of rail link between London's nearest, most prestigious, airport and the city centre. Yet it did; and even that was merely a spur added to the Underground's Piccadilly Line, with frequent stops, opened in late 1977. A dedicated Heathrow Express railway line to Paddington was inaugurated only in 1998, over half a century after this ever-expanding 'great world travel centre' had first opened for business.

were bussed to their flight in those days. Most simply walked a few yards out to the aircraft from the terminal door). The great majority are Vickers Viscounts or Bristol Britannias. This is not surprising, given that by then the Viscount was the mainstay of BEA's fleet and Bristol's new airliner would, it hoped, shortly become the same for BOAC on its long-distance routes. In addition, there are glimpses of some triple-finned, twin-engined Airspeed Ambassadors. These excellent aircraft were built by the company Nevil Shute had co-founded and were named 'Elizabethan Class' by BEA in honour of the new Queen. (It was one of these, G-ALZU, that crashed with the Manchester United team in Munich.) Also identifiable in these pictures are several non-British aircraft: BEA DC-3s, for instance; a Quantas Super Constellation, a BOAC Boeing 377 Stratocruiser with its distinctive 'double bubble' fuselage and one of Sabena's twin-engined Convair Metropolitans. But the overall impression is of a predominance of all-British airliners on their home ground. That this was intentional is clear from the first page, where readers are urged to 'fly British by BEA or BOAC'. Beneath a Union Jack waving across the top of the page are photos of a Viscount and a Britannia with the rhetorical question '2 of a kind?' answered immediately by the text below:

Yes, they're both British for a start. The propellers of both are powered by gas-turbine engines. Both get top marks for streamlined design, fast flying, elegant and comfortable cabins. Now for their differences. One is the magnificent BEA Viscount – the most famous airliner flying on European routes. The other is the new BOAC Britannia – the most advanced turbo-prop airliner to fly on the longer international routes. One final word: about the passengers who fly in these two airliners. They get the finest service in air travel today; service you only get when you fly British, by BEA or BOAC.

Exactly what kind of service is presumably illustrated on the inside cover opposite, a colour painting of a stewardess offering a bronzed and beaming first-class passenger a chromed salver on which there rests a packet of cigarettes ('You asked for Benson & Hedges cigarettes, Sir'). The stewardess is obsequiously inclined like a dutiful butler. Her blue military-style uniform with flap pockets makes her look like a WAAF whose duty it recently was to bring officers their tea as they planned that night's raid on Bremerhaven. Behind her is a view of the cloud-scape beneath the aircraft's port wing through a plate glass window of a size more suited to a railway carriage, fancifully at variance with the comparatively small windows in aircraft of the period.

Among the other things one notices about this booklet's glimpse of the future Heathrow is that all the aircraft depicted are propeller-driven. The Viscounts and Britannias simply suggest the dominance of the new generation of British turboprop engines over the conventional American piston-engined Stratocruisers and Super Constellations. There is no sign of a jet aircraft anywhere. The Boeing 707 was then still more than two years away from PanAm's inaugural New York–London flight in October 1958. Above all, there is no mention of the Comet: a sadly suggestive absence. Had this booklet been published three years earlier it would have been stuffed with seductive pictures of the world's first jet airliner, each one captioned with drum-banging prose. But the Comet was grounded and London Airport in 1956 had fallen out of the Jet Age, even though the turboprop Viscounts and Britannias were leaving behind them that new, reassuringly futuristic reek of burnt kerosene.

The Viscount was a lovely little airliner. Maybe after all it justified British hubris in taking on the United States at their own game. It was the first of a small but impressive handful of first-rate passenger aircraft the post-war British industry managed to produce. Certainly it was the only one that was a commercial success and paid its way – nearly 450 being built and sold around the world to sixty different operators. The others all fell victim to the British disease of becoming hopelessly delayed and arriving too late to make the commercial impact they deserved. The Viscount began as yet another response to the Brabazon Committee's wartime call for a medium-haul aircraft to challenge the foreseen American monopoly of commercial aviation after hostilities ended. One result was Airspeed's conventionally powered Elizabethan; the other was Vickers's turboprop Viscount. In its early development the Viscount prototype was known as the Viceroy; but after India's independence in 1947 it was obvious that this would not do. Vickers, set on coming up with suitably noble names beginning with 'V', clearly hadn't seen that 'Viceroy' might embarrassingly echo Louis Mountbatten's newly defunct title. Flicking through the dictionary to find a substitute they came up with 'Viscount'. This represented several ranks' demotion in the nobility stakes and was still not a great choice because no one had reckoned on Americans not knowing how to pronounce it. When the first Viscounts were sold to the US, Capital Airlines had to issue their staff with stickers saying 'PRONOUNCED VICOUNT'.[3]

The Viscount was the world's first turboprop-powered passenger aircraft, and it brilliantly challenged the American hegemony by being faster, smoother, quieter and more economical than all comparable US commercial aircraft of the period. It quickly made a name for itself by exemplifying the advantages of the new turbine-driven engines that represented a sort of halfway stage between the old piston engine and a full-blown jet. Because these engines ran on kerosene rather than aviation spirit they offered a greatly reduced fire hazard. To realise its turboprops' full economic advantage the Viscount cruised best at 25,000 ft or slightly higher, which required a properly pressurised cabin. This was more expensive to manufacture but well worth it in terms of fuel economy as well as passenger comfort because the Viscount could usually fly above the worst of the weather. The smoothness of its flight was frequently demonstrated by passengers eager to test advertisers' claims by using their seat-back tables to balance coins on their edges.

My first journey in a BEA Viscount in 1959 (for ludicrous and foredoomed amatory reasons) was from what had by then become Heathrow. My flight to Gibraltar was also my first in a turboprop aircraft. From its earliest appearance I had been captivated by the aircraft's neat looks. I liked the tailplanes' steep dihedral and I also very much liked the styling around the cockpit that seemed to tuck it between little streamlined cheeks. What appealed above all was the slimness of the four engine nacelles protruding far in front of the wings' leading edges. Here were no clunky great Wasp radials of the American airliners but slender Rolls-Royce Darts resembling jets with four-bladed propellers. Before going aboard I ducked beneath the wing and peered into the sooty ovals from which the turbines' hot gases would stream: jet orifices of a sort and reeking of paraffin-fuelled power. A Viscount of an earlier flight had started up and taken off with that thrilling sound – by now familiar from a dozen air shows – of the Darts' rising whine, eventually settling to a muffled scream chopped into fragments by the whirling propellers: the auditory equivalent of strobe lighting. My seat was over the starboard wing and I had an excellent view of the long polished nacelles and the ground crews' NO STEP warnings stencilled in black on the unpainted metal wing.

I watched start-up entranced, seeing how the propellers' flickering arcs changed their shade of grey as the pitch of the blades altered from 'feather' through 'coarse' to 'fine'. The doors thudded shut, cutting off the engines' shrill frequencies. The old magic of that childhood Dragon Rapide flight returned, where everything involved in flying seemed to shine with a lacquer of significance and become exceptional. The miracle of inducing transparent air to support thirty-three tons of metal and fuel and flesh as though they were feathers coated each humble piece of hardware involved and even lent a mystique to the ground crewman pulling the yellow starboard chock clear and walking off, dragging it behind him, before turning to give the all-clear signal to the pilot. Even the blind rivets on the wing were a source of fascination as we taxied out to the end of the runway, turned with a blast from the two outside engines and went straight into take-off with a surge of full power that brought a rush of excitement. (How naive this all sounds half a century later, when many people fly more frequently than they travel by bus, and with scarcely a thought for their commonplace conveyances!) Even at that time the Viscount's extraordinary smoothness was becoming old hat to seasoned travellers, for by now the big transatlantic jets like the Boeing 707 and Douglas DC-8 were in regular service, not to mention the Comet 4, and passengers were beginning to take for granted the smoothness of jet-powered flight. Today, nobody any longer remarks on it because they have never experienced the vibration and din of piston-engined aircraft. The Viscount had offered relief from this as early as 1950 when it entered service with BEA. It is not surprising that it sold around the world as a mainstay short- and medium-route workhorse. Even by 1956 it had earned Vickers approximately £62.5 million in exports, equivalent to some £1.25 billion at today's values, of which £40 million came from dollar orders.[4] It continued in service with BEA well after that airline and BOAC had merged in 1974 to form British Airways, flying until 1985 when the fleet was sold to charter operators. One or two Viscounts are believed still to be flying regularly in Africa. It was Britain's most successful commercial aircraft and its clean elegance could turn heads at air shows even today.

In 1960 the British Aircraft Corporation was formed from Hunting, Vickers-Armstrong, Bristol and English

Electric. It inherited the original design of its BAC 1-11 from Hunting Aircraft. Rear-engined like Sud Aviation's pioneering little Caravelle, it was intended as a short-haul jet replacement for the Viscount. Unfortunately, ever-changing requirements by BEA meant it had constantly to be redesigned to accommodate more and more passengers, and these cumulative changes badly delayed an aircraft that should have reached the market much earlier. It was also inauspicious that in October 1963 the prototype crashed, killing all on board including the pilot, the erstwhile speed record-holder Mike Lithgow. The reason for the crash was a deep stall: exactly what had killed Bill Waterton's deputy, Peter Lawrence, back in 1953. Like his Gloster Javelin, the BAC 1-11 also had a T-tail with the tailplanes mounted on top of the fin. BAC designed a stick-shaker (another first) that did as the name suggests, shaking the pilot's stick to warn of an incipient stall – a device now universally used. That problem cured, the 1-11 was an excellent short-haul jet and, since despite delays it still managed to anticipate its chief rival, the DC-9, it was a reasonable commercial success, even being bought by some US airlines. A total of 244 were built, several under licence in Romania. The aircraft's ageing Spey engines caused it to fall increasingly foul of noise restrictions, yet the UK's last operational 1-11 flew on until the end of 2012.

Another short-haul jet emerged at much the same time. This was the Trident, the Hawker Siddeley HS.121 (before the merger de Havilland's D.H.121). It was the world's first tri-jet airliner as well as the first able to make completely blind landings. Again, this was an aircraft whose potential was fatally undercut, first by BEA shilly-shallying and then by the managerial chaos created by the government-enforced post-Sandys mergers. Consequently the Trident came into service after its chief rival, the Boeing 727, and only 117 were built. It deserved far better.

The aircraft that was intended to be even more successful than the Viscount was the Bristol Britannia. After the Brabazon's failure Bristol were pinning their hopes on its more realistically sized successor to capture much the same long-distance market as the Comet, but with the much greater fuel economy offered by turboprops. The public had first seen the Britannia when 'Bill' Pegg, Bristol's chief test pilot, showed the prototype somewhat gingerly at Farnborough in 1952, not long after Geoffrey Tyson had performed with the massive Princess flying boat. Pegg was cautious because although he had first flown G-ALBO nearly thirteen months previously, it had been plagued with landing-gear problems, the last of which had occurred only a fortnight before the SBAC show. On the day, G-ALBO performed perfectly. It was already painted in BOAC's colours in honour of its launch customer and its great white tail fin with the Union Jack decal on the rudder gave a great impression of calm steadiness of purpose. (The huge rudder was at least partly intended to counteract any sideways swing if an outboard engine failed.)

Seventeen months later in February 1954 Pegg had a less satisfactory flight in the second prototype, G-ALRX. The landing-gear problems had long been solved but the new Proteus engines had given a good deal of trouble although this, too, seemed to have been overcome. On board that day were eleven others, including some heavyweight observers. Invited were Stanley Hooker, Bristol Engine Division's celebrated chief engineer; the Britannia's chief designer, Archibald Russell; and representatives from a keen potential airline customer, KLM. That morning Pegg lifted G-ALRX smoothly off Filton's long runway and headed north towards Herefordshire. They were still climbing when after seven minutes the oil temperature of no. 3 engine rose alarmingly and Pegg shut it down, ostensibly to demonstrate to his Dutch colleague (who was co-pilot in the right-hand seat) that the Britannia could climb perfectly well on three engines.

After a while the oil temperature in no. 3 was back to normal so Pegg restarted it and climbed through 10,000 ft when without warning there was a loud explosion in the same engine and almost immediately the fire warning light came on and the alarm sounded. In swift movements he at once pulled back that engine's green-topped throttle lever before he reached over to the Engine Fire Control panel and flicked no. 3's vertical row of toggle switches down: alarm, fire doors shut, oil cocks shut, low-pressure fuel off, high-pressure fuel off. The co-pilot looked back through the window and reported fire, and at that moment Archibald Russell appeared on the flight deck to report a 'hell of a fire' in no. 3 with flames streaming back as far as the tail. In fact what had happened was that the reduction gear in that engine had stripped its cogs. Nearly all propeller aero-engines needed a system of fixed gearing to reduce the speed of the propeller. This was because the speed at which the engine was designed to run most efficiently would, if it drove the propeller directly, turn it much too fast (at only a little over 1,800 rpm the tips of a 12-foot diameter propeller reach the speed of sound, which is not only intolerably noisy but very inefficient). Hence the Britannia's Proteus engines incorporated reduction gearing immediately behind the propeller. On this occasion a fault had caused these cogs to shatter, destroying the drive to the propeller, and the suddenly unburdened no. 3 engine had raced itself to destruction, sending a shrapnel-like burst of hot metal fragments into the oil tank which had promptly ignited. The engine's extinguisher system seemed to be having no effect on the fire. Suddenly, this was a critical situation. They were at 10,000 ft with a raging fire in a wing that contained 2,000 gallons of fuel. Pegg shut down no. 4 engine as well and now had to take a quick decision on which twelve lives would depend. Should he try to return to Filton or attempt a crash landing? They were now over the Welsh mountains, so this last hardly seemed an option. He turned back on a Bristol heading but in everybody's mind was the knowledge that with the slipstream fanning the fire to white heat it was only a matter of time before the starboard wing's main spar began to soften and the entire wing folded up. If that happened the aircraft and all aboard her were doomed. There would be no hope of recovery or survival.

Descending rapidly and with Bristol in sight, Pegg could see the Severn's mud flats glinting in the morning sun and realised that the tide was out. He immediately opted for a crash landing on the mud. With only the two port Proteus engines keeping the aircraft in the air, the advantage of the Britannia's big rudder in maintaining the aircraft's steerability became apparent. He was just lining up for a 200 mph wheels-up landing on the mud somewhere south of Sharpness when without warning even these two engines stopped. Worse still, Pegg calculated that the aircraft would now hit the ground at exactly the point where the channel of a small river cut across the beach into the estuary, probably deep enough to flip the Britannia on to its back or send them cartwheeling across the flats in flames. Working frantically, the two Bristol engine technicians on the flight deck managed to reignite both port Proteuses and at the last moment the Britannia regained enough flying speed to avoid falling like a sack of dumbbells. Shouting to everyone to brace, Pegg brought the aircraft with flames and thick black smoke trailing from its starboard wing down on its belly in the glistening mud to a flawless crash landing. G-ALRX slid 400 yards with uncanny smoothness until the last moment when it turned abruptly right towards the sea and stopped. With yet another stroke of luck the mud seemed to have put the fire out and the entire party was able to scramble out uninjured before wading through the silt to safety. It was a tribute not only to first-rate piloting but to an immensely strong airframe.

Gradually the Britannia surmounted its problems so that by 1956 my *London Airport* booklet boasts a confident, full-page advertisement for 'The Whispering Giant'. Above a night photo of G-ANBJ on the tarmac the copy reads: 'The Bristol Britannia is the largest and most advanced airliner flying. Capable of carrying up to 133 passengers with all the smoothness, comfort and quietness associated with turbined-powered [sic] flight, this eighty ton aircraft cruises at 400 mph over ranges from 6000 down to 200 miles more economically than any other airliner.' An insider's view of this aircraft's development, as of the company that built it, was given by Sir Peter Masefield, who had been running BEA before joining Bristol Aircraft as their managing director at almost the exact moment this advertiser's copy was being written. It is instructive because it reveals much about the way in which the chances of the Britannia's success – like those of so many other British aircraft of the period – were squandered. Masefield was a highly experienced pilot before he was a businessman, and had no-nonsense views to match.

Almost as soon as he had sat down in his office he fell foul of the clannish way in which this family-run business operated. Thereafter there were frequent rows whenever he made any sort of public announcement without first clearing it with the directors. Worse, the business procedures he had put in place to run BEA profitably were clearly all wrong for Bristol. 'I was made to understand that [the company] was a place for gentlemen, who were above such things as balance sheets.'[5] Being above such things as balance sheets was not likely to prove a successful way to run a business enterprise, and so it proved at Bristol despite the company having some excellent designers, brilliant engine builders, and an impressive list of (far too many) projects including military aircraft, helicopters and guided missiles. Almost as soon as he had accepted the post at Bristol Masefield was offered the managing directorship of BOAC, the British Overseas Airways Corporation, the national carrier on routes outside

Europe. Had he taken the job British civil aviation might have developed along more rational lines. But he honoured – with some misgivings – his prior decision because it was 'a challenge'. This challenge was, in mid-1955, to make Bristol's turboprop Britannia competitive with first-generation US jet airliners such as Boeing's new 707 (which was even then being test flown) and Douglas's DC-8, both of which could cruise at least 100 mph faster than the British aircraft while carrying more passengers. To make things still more difficult Masefield was up against the awful slowness that pervaded Britain's aviation industry and which among Bristol's board had, in his own words, now attained a state of 'abysmal lethargy'. The Britannia's development had been dawdling along for eight years already and it still had not overcome its engine problems. 'All that this great aeroplane needed was engineering manpower,' Masefield lamented, '[but] from the start it had suffered from a total absence of the American spirit which simply trampled problems to death, usually in hours. At Bristol, problems simmered for months.'[6]

So far as the Britannia's engine problems were concerned, these had been simmering for years. The engines were Bristol's own Proteus turboprops, originally designed for the second Brabazon prototype, the first having been broken up in 1953 with fewer than 400 flying hours and not a single customer, leaving behind as its two useful memorials at Filton a gigantic purpose-built hangar and the lengthened runway. The Brabazon had cost Bristol some £3.4 million to develop (over £73 million today) and the new Proteus engine must have looked like the only thing of value that might be salvaged from the entire project. Ten of them were already installed in the Saunders-Roe Princess flying boat in a complicated coupled arrangement that had already caused hideous engineering problems. By 1955 the engine was still giving trouble. Like Masefield, Bristol Engine Division's chief engineer, the great Stanley Hooker, was outspoken about the regime he served:

In those days I used to say that the biggest obstacle to Bristol's progress was a Bristol lunch. In each factory the top man had his own little private dining room. We would start with hot canapés while we partook of sherry. Then we would sit down to a multi-course lunch ending with cheese, fruit and coffee – and on occasions brandy. The whole lot would last from 12.30 until at least 2.30, about twice the time we took at Derby [i.e. at Rolls-Royce, where Hooker had previously worked].[7]

(This heavy lunching seems to have been standard practice at managerial level in much of the industry in those days, and even considerably later – we shall be meeting it again. Presumably the High Table or club ambience reminded the 'gentlemen' directors of their Oxbridge days and softened the indignity of having to think of themselves as industrialists or even – God forbid – 'in trade'. It was part of a nineteenth-century mindset that proved so disastrous to Britain in the competitive high-tech world of supersonics and mass travel.) The specific problem with the Britannia's Proteus engine that Hooker had to solve was an over-complex reverse-flow system that derived from its having been designed for the Brabazon and the Princess, both extremely large aircraft. Had he known they would be cancelled he might have simplified the engine's design, thereby avoiding the persistent icing problems the engine encountered that were to delay the Britannia's entry into service by a further two years.

Yet even that delay could have been avoided had BOAC really wished. Back in 1949 the airline had contracted to buy twenty-five Britannias, but since then it had constantly changed its mind about the number of passengers it expected them to carry, wavering between sixty-four and eighty-three, the upper figure of which required design changes that made the aircraft nearly five and a half tons heavier and threw out all Bristol's engineering and pricing calculations. It was true that BOAC found itself in an awkward position. It was Britain's major national carrier, and as such was expected to fly British aircraft for which it effectively paid subsidised prices; but it made no bones about wishing it could buy American even though the government had no dollars to spare. At the time this was decried by some as grossly unpatriotic, but with hindsight a case could be made in BOAC's favour for having decided early on that American domination of the commercial aviation scene was in the long run inevitable and unbeatable, and that it would much rather fly a uniform fleet of Boeings (or Douglases or Lockheeds) than a mixed bag of American and British aircraft, with the added complication and expense of different spares, equipment and qualified mechanics, all to be replicated worldwide.

The icing problem in the Proteus engines was not, in fact, very serious. It occurred only in easily avoided circumstances and Hooker engineered a series of remedies, any one of which worked reliably. It was thus not a major safety factor. But BOAC deliberately exploited the situation, announcing it in public as a deadly hazard – an extraordinary thing for any airline to do with an aircraft it was contracted to buy. Masefield was exasperated.

We continued making unnecessary modifications for *two years*. It delayed the Britannia's entry into service from March 1955 until February 1957 ... What upset me was BOAC's determination to rubbish in the most public way what might have been a world-beating all-British airliner ... The ultimate result was that the Britannia's acceptance by BOAC was repeatedly postponed until the American big jets were also on the point of entering service. It destroyed the Britannia's widespread market appeal, and the cash-flow crisis came within an ace of driving the Bristol Company bankrupt.[8]

In the meantime Hooker and Bristol Engine Division had come up with a superb new turboprop engine, the Orion, which when installed in a Super Britannia would give it a much higher cruising speed of 470 mph for the same range. 'It was obvious', Masefield wrote with new assurance, 'that we could compete with the 707 and the DC-8.' Given that the American jets could cruise over 100 mph faster with a larger payload, this seems optimistic to the point of wilful stupidity. It is hard to accept it as the considered position of a man who was not only an experienced pilot but a proven expert on the business of commercial aviation who had turned BEA's fortunes around. In 1956 he took the tenth production Britannia, G-ANBJ (the very aircraft photographed for *London Airport*), on a world sales tour, doing much of the flying himself. It aroused considerable interest even in the US.

He entered into serious negotiations with Eastern Airlines, and then later with TWA, whose eccentric president, Howard Hughes, test-flew the Britannia himself impressively well, having first removed his shoes. Yet both sales prospects foundered once it became clear that Bristol couldn't meet the orders by the dates required. The company lacked the capacity for mass production. Eventually, a total of eighty-seven Britannias was built, a mere fraction of what might have been. In 1957 one of Duncan Sandys's many cost-cutting measures was to stop supporting the Orion's development. This killed an engine of extraordinary promise stone dead, and with it all hope of a Super Britannia.

The Britannia's history is tragic in its revelation of how an excellent aircraft's prospects were steadily killed off by managerial and political incompetence. The aircraft's appearance in the advertisements and illustrations of the *London Airport* booklet of 1956 has an aura of retrospective melancholy, for even then the Whispering Giant was doomed to be superseded by the new American transatlantic jets within two years. Even so, it still became the first aircraft ever to carry 100 passengers non-stop both ways across the Atlantic, and it could fly non-stop from Tel Aviv to New York. It looked good, too: for such a large airliner it had extraordinarily graceful lines. There was no dihedral at all on any of the flying surfaces: together with its huge square-topped tail fin, everything contributed to an overall impression of *flatness*. If this flatness promised exceptional stability, other features suggested the Britannia was swift as well. The Proteus engines were comparatively slender, like the Viscount's much smaller Darts; they were also slightly staggered, the two outboard engines being set a little behind the inboard. This was because the leading edge of the wing had a distinct degree of sweepback, especially compared with its four-motor propeller-driven Douglas and Boeing contemporaries. The Britannia's linearity was increased by its narrow nose and rakishly paned cockpit. In its white-top BOAC 'Speedbird' livery with the wings and underside left as polished metal it gave off an air of eager capability. Despite the insuperable handicap of its late arrival, this fondly remembered aircraft still saw respectable service with several airlines as well as with RAF Transport Command, the last examples disappearing in the eighties. It also afforded me one of the more memorable flights of a travelling lifetime when in 1968 I was offered a free trip to Hong Kong aboard a Lloyd International Airways cargo Britannia out of Stansted.

I and four others had been instructed to drive on to the airfield by a back gate and go straight to the aircraft out on the pan. In those days Stansted was a lot smaller and so informal that there was no one at the gate to ask, so we had to drive around the flight line a bit before we identified our particular aircraft. We wandered up the steps and introduced ourselves to the affable crew (all ex-RAF), who were busy with pre-flight checklists and lading manifests. We soon discovered there were no seats: the entire fuselage was stuffed with bales of cotton material. The deal was that we were to fly out to Hong Kong where there would be a three-day turn-round, in which time the city's sweatshops would convert our cotton bales into thousands of pairs of knickers for Marks & Spencer. These would then be reloaded and we would return to the UK. Just before we took off somebody remembered to ask whether we had brought our passports with us, otherwise 'the chaps in Honkers might get a bit stroppy'. We wedged ourselves on top of the bales in the four-foot crawlway beneath the top of the fuselage and so began our sedate passage to the other side of the world.

It was extremely comfortable, lying Princess-and-pea-style atop a vast bed of cotton six feet deep; but the journey seemed to drag even though we made good time at a steady 400 mph. We ate ageing British Rail-type sandwiches that curled more the further east we flew. Occasionally we crawled down towards the tail where there was a sort of cupboard with an Elsan chemical toilet bolted to the floor. We made refuelling stops at airfields like Dum Dum outside Calcutta where the Indian authorities obliged us to disembark but would not allow us to leave the aircraft's shade. We cowered from the sun beneath a wing while the fuel bowsers mustered with their leaky hoses. The puddles of kerosene spread until by the time we finally re-embarked our shoes were reeking with AVTUR (aviation turbine fuel). This wait gave me plenty of opportunity to examine the aircraft's landing gear at leisure and otherwise poke and pry, dabbing drops of hydraulic fluid from concealed nipples with my fingers and smelling them. Since the Britannia was an altogether larger and taller aircraft than the Viscount I couldn't sniff at the Proteus exhausts because the jet effluxes vented well out of reach, above the wings' trailing edges rather than below.

The somewhat bored crew were wonderfully indulgent of my curiosity, and from Dum Dum onwards I spent most of the journey in the cockpit, which with its quilted roof seemed both luxurious and businesslike. Trying not to sound like a twelve-year-old, I plied them with endless questions which they seemed happy to answer, as if pleased to find a fellow enthusiast. Particularly at night, the 'office' felt like a natural home: the bright constellations overhead, the orderly rows of unwinking panel lights, the long introspective silence above the engines' unwavering hum and the radio officer's occasional muttered conversations behind us as he spoke to traffic controllers somewhere in the void beneath. At the time, the man in the right-hand seat had control and he eased back a little on the control column with its distinctive horn-type grips like a splayed-out 'M'. We gained some more altitude before asking for and getting permission to overfly South Vietnam, in that year of the Tet Offensive an intense war zone. We watched the wandering lightning flashes of a tropical storm silently pooling in hectic pearlescent patches under the roof of clouds far below us and it was easy to pretend that a great battle was raging on the ground. And in due course, long after sunrise, I was allowed to perch on the jump seat behind the pilot's and co-pilot's seats when landing at Kai Tak ('If you're going to lurch forward, for Christ's sake don't bang these throttles, OK? We'd be a bit buggered'). Thus I was afforded an unforgettable airman's view of the hair-raisingly steep approach over high-rise buildings, a last-minute right-angled turn before touchdown, full reverse on the propellers and braking to a smoking halt a few yards short of the South China Sea.

Three days later the pilots, much the worse for floozies and drink, yawningly seated themselves once more and took off, rotating rather too abruptly and banging the underside of the tail on the tarmac ('Oh, well done, Geoff. I bet that's spilt the sodding Elsan'). Back at Dum Dum for our first fuel stop we checked for damage but there was nothing beyond some scraped aluminium skin. Those were the days when flying was still an adventure, informal and friendly enough to afford huge and unexpected pleasures. By the end of the week I had a working knowledge of the Britannia's cockpit layout and had been instructed how to wind on flaps or trim, to start and stop the engines, lower and raise the undercarriage and steer the nose wheel using the recessed tiller by the pilot's left hand. All unbeknownst to me then, fate had scheduled me to begin flying lessons in Manaus, Brazil, later that year; but those two thirty-six-hour Lloyd International flights in the Britannia afforded me almost as much instruction and pure enjoyment as finally getting my hands on some real live controls. What madeleines once did for Proust would today be done for me by the scent of fresh cotton knickers and AVTUR, with maybe just a hint of Elsan fluid.

Peter Masefield was still managing director at Bristol when Duncan Sandys's White Paper was published in 1957. He was every bit as scandalised by its consequences for commercial aircraft as were the military pilots left glumly surveying the prospect of the RAF's disbanding in favour of guided missiles. 'Thanks to Government (i.e. Sandys) policy, we were entering a long period of Cloud-cuckoo-land in which virtually the entire energy of the Boards of the British aircraft companies was devoted to organising partnerships ... In effect, the world-class British aircraft industry was being deliberately destroyed.'[9] Good rousing stuff; except that where making commercial aircraft was concerned the British aircraft industry was manifestly *not* world class, although as we know it was quite capable of coming up with an occasional aircraft that would potentially have been world class had it not been for the foot-dragging incompetence of its own manufacturers, the mulish resistance of the nation's major airline, and the whimsical mutability of government policy. Of the fifty or so different British civil aircraft designs that actually managed to reach production, all but half a dozen were never built in significant figures; and with the sole exception of the Viscount even those figures were nothing like significant enough to save the industry.

The sort of thing a potential world-class airliner was up against in Britain can be seen in the case of Vickers's successor project to the Viscount, the V.1000. In 1951 the Air Ministry had issued specifications for a long-range strategic transport for the RAF to replace their fleet of propeller-driven Handley Page Hastings. Since the new transport would have to service the needs of Britain's V-bombers then under development and which might be deployed anywhere around the world, it was made clear that it would have to be jet-powered, capable of at least 600 mph and with great range. One of the money-saving stipulations was that it must be based on an existing airframe. All four V-bomber manufacturers (Shorts, Vickers, Handley Page and Avro) came up with variations of their warplanes, while de Havilland offered a stretched version of the Comet. Of the designs submitted, the one by Vickers was chosen: a much larger development of the Valiant with a good few changes, especially to the tail. In 1953 a contract was signed for a single prototype. So far, so good.

Meanwhile, it was obvious that this new transport could easily be configured for commercial use to become the world's first really big jetliner for the North Atlantic and Empire routes. Soon, BOAC was taking part in detailed planning for a civil version which would be known as the VC7. This aircraft would seat 100 passengers six abreast. Once the prototype V.1000 had flown the VC7 could follow swiftly on. By the end of 1955 the first V.1000 under construction at Vickers's Wisley plant was within months of completion when BOAC decided that its increasing weight would require the Rolls-Royce Conway engines to be uprated and concluded this wasn't feasible. Simultaneously the RAF's budget was drastically cut and on reconsideration it accepted it would have to forgo its new jet transport after all and make do instead with militarised versions of the Comet and the Britannia. It pulled out of the project and BOAC promptly followed suit, a ministerial spokesman saying the airline was 'satisfied that it can hold its own commercially on the North Atlantic route until well into the sixties with the Comet 4 and the long-range Britannia'.[10] That November the V.1000 was cancelled, and with it the VC7. Incredibly, less than a year later the House of Commons was informed that permission had been given for BOAC to buy fifteen Boeing 707s costing £44 million 'in order that the Corporation may hold their competitive position on the North Atlantic route from 1959 to the 1960s. At that time no suitable new British aircraft can be made available for that purpose.'[11]

In keeping with ruthless tradition, the unfinished prototype in Wisley's hangar was broken up and all the jigs destroyed. George Edwards, Vickers's brilliant chief designer who had produced the Viscount and the Valiant and whose baby the V.1000 was, commented grimly, 'It's no use leaving a corpse about for the chaps to mourn over.' The realpolitik of British aviation. BOAC's extraordinary volte-face was made all the more grotesque when it turned out that the Boeing 707s it had ordered were to be powered by uprated Conways, the very engine it had said couldn't be developed to power the VC7. Such was the end of what had promised to be the first-ever jet airliner capable of flying the Atlantic non-stop with a payload of 100 passengers, and all a good two years before its potential American rivals. Later, George Edwards described the whole miserable affair as 'the biggest blunder of all'. He would say that, wouldn't he, given that he had designed the aircraft; but it's hard to disagree. The way the entire matter was handled, right from the MoD's issue of an Operational Requirement through to BOAC's calculated duplicity, typified the way one of Britain's truly great and inventive aircraft companies could be reduced to a pawn in a game played offstage by Civil Service advisers, government ministers, Treasury officials and the airlines. By the end Vickers were left having to abort the new baby taking shape in their sheds and deal with a suddenly bereft and discouraged workforce. It was clear that, despite its initial public enthusiasm for the VC7, BOAC was always likely to be swayed by its pro-American faction, and so it proved.

Their case was too strong, able as they were to point to the not-yet-flown Comet 4 and the slow-selling Britannia – both of which aircraft they had contracted to buy. Having to add the still-unbuilt VC7 was too much, so they turned and ran in the only direction they had ever really wanted to go.

The airline's behaviour was even more peculiar because they must have known perfectly well that while the Boeing 707 was excellent for the transatlantic route, it was less suited to the difficult 'hot-and-high' airfields of East Africa, the Far East and elsewhere. American airliners were designed for the very long runways typical of their spacious native land. Most couldn't operate from comparatively short strips, least of all where the air was hot and thin and afforded less lift. As the national carrier, BOAC had quite a few hot-and-high airports on its so-called Empire routes and in 1957, barely a year after summarily rejecting the V.1000, it shamelessly returned to Weybridge to suggest Vickers now build them a new long-range high-performance jet airliner with 'hot-and-high' capability. By then Vickers were coming to the end of building Valiants and were beginning to worry about having enough work in the future. This new airliner, BOAC explained, would be a private venture – meaning that it would be a big financial risk for Vickers. Not only would the order be small, but after its recent treatment by the airline Vickers placed little faith in its word, especially as there would be no financial backing from the government. Yet despite everything they went ahead, and the VC10 was born.

The trick to getting a large airliner to unstick from short tropical runways or land on them at a slow enough speed was to have masses of power and nothing but flaps and slats at the wings' trailing edges to ensure lift. The question then became, where to put the engines? Ever since the Boeing B-47 jet bomber of the late forties, American designers had pioneered hanging the engines on pods beneath the wings. There was obviously a penalty in drag and it could sometimes slightly restrict the control surfaces at the wings' trailing edges, but none of this mattered too much as long as there was plenty of runway available. But being sited low, podded engines have the disadvantage that they more easily pick up any stones or other debris on a runway, with potentially disastrous results. (This is known in the trade as FOD or 'foreign object damage'.) And hot-and-high airfields in the more far-flung parts of Africa and the Far East often had runways that between flights – which might not even occur daily – were home to herds of goats and were played on by schoolchildren, so could be littered with all sorts of rubbish. In due course podded engines became the basic configuration for practically all big American jets, from the B-52 to the 707 to the 747 'Jumbo' and beyond. Today, most airports have goat-free runways of ample length, which partly explains why this type of design, though aesthetically clumsy, has become standard in both Boeings and Airbus aircraft (a further reason being that podded engines are the most accessible for servicing). But back in 1957 Vickers decided that for good reliable hot-and-high performance it needed an aircraft with clean wings, maximum control surfaces and high-mounted engines. So when George Edwards drew the VC10 he hung the four Rolls-Royce Conways at the back, as the French had so enterprisingly pioneered with their much smaller twin-jet Caravelle.

Over the next few years the project encountered all too familiar setbacks. As usual, BOAC kept changing their mind about both the specifications and the number of aircraft they wanted to order, and things only began to look better for Vickers when the RAF began taking an interest in the VC10 as the fast, long-range transport they still needed. When Jock Bryce and Brian Trubshaw took the prototype up on its maiden flight in June 1962 they did it from Brooklands, whose runway was a mere 1,300 yards long. This was sheer chutzpah on its designer's part. Edwards later remarked, 'Everybody thought I was bonkers to make a first flight with a big new-style aeroplane like that out of a little saucer of an aerodrome like Brooklands.'[12] But he knew he had designed what was then the most powerful airliner in the world, and he knew it had real short-field performance, and he was right. It flew beautifully, taking off and landing at speeds well over 20 mph slower than a Boeing 707.

In its testing over the next year the VC10 met with and overcame some initial stalling and drag problems before going into service with BOAC in 1963, becoming instantly popular with both pilots and passengers – the latter particularly liking the quietness associated with having the engines at the back. It was a beautiful aircraft to look at, with its long slender nose and big fin topped by sculpted, almost *windswept* tailplanes headed by the characteristic 'bullet' fairing containing such things as the instrument landing system aerials. In its BOAC livery it looked outstandingly elegant with its clean, uncluttered wings, making a sharp aesthetic contrast with the 707s and DC-8s bristling with engine pods. (Probably the only aircraft that has ever managed to make jet pods look sleekly part of the overall design was Convair's snazzy fifties delta bomber, the B-58 Hustler.) What was more, the VC10 was soon getting higher payloads than its American rivals. But this seemed to cut no ice with BOAC's new chairman, Sir Giles Guthrie. Reverting to type, he suddenly decided to convert the airline back to an all-American fleet. He cut the order for thirty Super VC10s to seven, then cancelled the lot. His airline's original stipulations about range and take-off distance had produced an aircraft that had a higher seat-to-mile cost than the 707; but this would always have been true. In aircraft design, as elsewhere, there's always a price attached to increased performance. Yet again history repeated itself as BOAC 'orchestrated a campaign against the VC10'. As George Edwards once more bitterly observed, 'if you couldn't sell [an aircraft] to your own airline the chance of selling it outside wasn't very great . . . It was a sorry story. The whole relationship between BOAC and the industry at that time was bloody awful and did the industry a power of no good.'[13] To add further irony, it turned out that in the long run the comparatively few VC10s that flew proved more profitable than the Boeing 707. Charles Gardner summed it up: 'So the Super VC10, denigrated in advance by BOAC as too expensive to operate economically – and for which they obtained some £30 million in subsidy as recompense – turned out to be actually cheaper to fly than the 707 and also to attract more passengers.'[14]

Even so, the business of attracting passengers did depend on the airline having done its market research. When I first flew on a BOAC VC10 to Tripoli in 1965 the aircraft was full. A few of us disembarked – perhaps twenty-five – and as far as I remember we seemed to be replaced by others for the next leg of the journey to somewhere like Khartoum. But a few years later I boarded a British Caledonian VC10 in Recife, Brazil, bound for London Gatwick. I remember BCal fondly as a thoroughly splendid airline to fly with, but on this route they can't have done their sums too well because I was one of only eleven passengers rattling around in this large aircraft. (This must have been before 1972 because in that year BCal replaced its four VC10s on the South American run with Boeing 707s.) It was a night flight, and I remember the other passengers as elderly and falling asleep as soon as our dinner had been served. After that I went forward through a desert of empty seat rows to a galley to chat with the air hostesses in their tartan finery, and bit by bit wangled an invitation on to the flight deck.

Thereafter I spent almost the entire flight in the cockpit, avid as ever for technical detail. Once again the crew seemed to be ex-RAF and only too happy to indulge me. One thing made enough impression for me to have already described it elsewhere.[15] VC10s carried a navigator who sat behind the pilots on the right in a sideways-placed seat at a work station facing its own oval window. Ours had been getting regular radio fixes from Dakar and Cape Verde far ahead in the darkness. Somewhere in mid-Atlantic he stood up and opened a panel in the cockpit roof. Here there was a valve-like aperture through which he thrust a periscope sextant and shot the stars to confirm our position. 'We like to make sure,' he explained. 'Back-up, really. Radio fixes are fine but it's nice to get confirmation from a completely different source.' The entire flight seems to me now to have been enchanted (youth and that); but some of that quality must derive from the way it encapsulated so much that has vanished – BCal, for a start, as well as those relaxed days before the world surrendered to abject paranoia about terrorism and it was still possible for a passenger to pass the night chatting in an airliner's cockpit. Nowadays, of course, airliners no longer carry navigators: the last went out with the VC10 and the Boeing 707. Most aircrew have probably never seen a sextant, let alone know how to use one. Navigation is increasingly done by GNSS (Global Navigation Satellite System). Old skills becoming extinct are somehow more lamentable, although less poignant, than the passing of aircraft like the VC10.

The aircraft surely had its faults. Its Conway engines were the world's first by-pass jet engines to go into service and they were thirsty and noisy. Although it was quiet for passengers, all of whom sat ahead of the engines, the VC10 would not qualify to operate out of today's major airports, nearly all of which have noise restrictions. But then it is getting on for fifty years old, although still looking as elegant and fresh as on its first roll-out. When it first flew it was a tour de force of design, particularly in its flight-control system and cockpit technology. For its December 2003 issue *Aeroplane* polled hundreds of pilots and listed their top ten most 'pilot-thrilling' aircraft. The VC10 came top, beating even the F-86 Sabre into second place. (The list is also interesting for including the Hurricane and not the Spitfire. The Hurricane is ranked no. 8, ahead of the P-51 Mustang at no. 9, the list being completed by the Lancaster – surely the most biddable bomber ever to fly.) As though to confirm the point, the story is told of the Vickers test pilot Desmond 'Dizzy' Addicott having to ferry a VC10 back from America in the sixties. It was empty but for its crew. Somewhere over the Atlantic on a whim Addicott instructed his colleagues on the flight deck to hang on and promptly barrel-rolled it. Unfortunately he had forgotten about the stewardesses down in the aft galley, who were alarmed to glimpse the ocean rear up and soar overhead before descending on the other side, now rather nearer. It must have been one of the later examples of old-fashioned aviator's exuberance in a supposedly staid and graceful airliner. It was one thing for Roly Falk to roll a Vulcan – it was, after all, a solid delta-shaped mass of muscle. But to roll an airliner the size of the VC10 showed not only high spirits but total faith in the capabilities of the designer and company the pilot worked for. Even today a VC10 still holds the record for the fastest crossing of the Atlantic by a jet airliner other than Concorde.

Until December 2013 you could still go to Brize Norton and see 101 Squadron flying the last VC10 K3 tankers. Standing by the fence at the end of Runway Two-Six you could watch one of these pale ghosts in camouflage grey taxi out towards you and turn on the numbers. Then the four Conways opened up and all forty tons of thrust thundered back over the blast pads' yellow chevrons.

Revelling once more in the friendly incense of burnt kerosene and hearing again that ripping crackle of sound was to fall prey to a deep nostalgia. Yet the sonic drubbing also bullied out a kind of unashamed pride: that this still-glamorous example of British engineering genius had racked up half a century of service. The sight of the Conways' splendidly flagrant carbon footprints disappearing into the grey cloud base while still battering Oxfordshire with their exuberant decibels – a sound so vigorous it could surely only be youthful – carried mournful overtones nevertheless. As with Vulcan XH558, it spoke so eloquently of everything this nation could so recently do, and of the men who so bravely and brilliantly did it in a time that has gone, and which can only come back to us now in remembered waves fading on the wind.

Fighter Jock Heaven

As we know, despite its world air-speed record Fairey's pretty and potent experimental delta, the F.D.2, never was developed into what could have been Britain's first truly supersonic fighter. Yet English Electric's response to the same Ministry of Supply requirement, the Lightning prototype P.1A, had already flown and was about to take on that role for itself. After the worldwide success of his Canberra, 'Teddy' Petter had turned his attention to building an interceptor for the RAF that would easily exceed the planned performance of the ambitious and ill-fated Miles M.52 project whose cancellation had so badly set back British prospects for true supersonic research. He completed most of the initial design work on the Lightning before leaving English Electric to go to Folland for what turned out to be an idiosyncratic – even ideological – belief. He considered that these ever bigger and heavier fighters were a mistake: that with smaller and more powerful jet engines becoming available the future lay in much cheaper fighter aircraft that could win big sales in developing countries with simpler needs. Once at Folland he swiftly designed the tiny Midge experimental jet fighter, which was soon followed by the very slightly larger Gnat. In due course the Gnat sold to India and Finland as well as to the RAF as a trainer, where it became familiar to countless British air show spectators for almost two decades in the hands of the Red Arrows aerobatic teams.[*]

The Lightning that Petter had designed before he left English Electric was astonishingly original in all sorts of ways, especially given the contemporary fascination with deltas. His aircraft's wings were narrow rectangles with a steep sixty-degree sweepback, cut off at the ends at right angles to the fuselage. It was as though Petter had taken a delta and removed a triangular chunk from its base. The proportionally tall, slab-sided fuselage was just as distinctive. Inside it, in an arrangement unique to the Lightning, he stacked the P.1A prototype's two Bristol Siddeley Sapphire engines one above the other, the upper engine staggered slightly behind the lower. These were fed from a single intake in the nose and the result was a great saving in drag since he could get twice the power of a single-engined fighter like the Hawker Hunter with a mere fifty per cent increase in frontal area. Apart from that, the eventual Lightning's Rolls-Royce Avons were much beefier than the earlier version of the Avon the Hunter used.

The result after development was a massively powerful, low-drag jet fighter that from the moment it entered RAF service in 1960 was recognised by pilots as the hottest ship around – and not merely in Britain. Better still: by a

---

[*] Petter's Gnat turned out to have other uses as well. In 1966 Donald Campbell's *Bluebird* speedboat was fitted with a Bristol Siddeley Orpheus jet engine from Gnat XM691, bought from the Air Ministry for £200. Campbell also fitted the Gnat's tail fin to his boat as a stabiliser. While attempting a record-breaking run on Coniston Water on 4 January 1967, Campbell and his teddy-bear mascot, Mr Whoppit, performed an unscheduled but almost perfect loop, unfortunately with fatal results for Campbell. Mr Whoppit survived, floating though sodden.

miracle of timing it managed to leapfrog Duncan Sandys's slash-and-burn approach to Britain's aerial defence, thereby becoming the nation's only home-built Mach 2-capable military aircraft. It remains the fastest and finest single-seat jet this country ever produced.

It was, of course, every red-blooded pilot's dream to fly the Lightning. It was only the third production fighter in the world to achieve Mach 2 (after Lockheed's F-104 Starfighter and – by only three weeks – Dassault's Mirage III). There was lively competition for a coveted place on the Lightning Training Course at RAF Binbrook in Lincolnshire, despite that station's notoriety for enjoying some of the most miserable weather conditions anywhere in the British Isles with regular rain, fog, gales and snow, as well as infamous days when it was claimed even the birds walked. Never mind: to fly the Lightning was the ultimate boy-racer's experience, 'passing the love of women', as one ex-Lightning pilot remarked, 'by a long, long way'. To be truthful, it was utterly terrifying as well as thrilling to think of climbing that ladder and worming awkwardly between it and the refuelling probe before squeezing yourself into the cramped cockpit ten feet off the ground. And there you'd be, in sole charge atop eighteen tons of brute muscle that as soon as the wheels left the runway could throw you on your back and explode into a near-vertical climb to where the sky began to darken. Yet Roland Beamont, arguably the luckiest dog of all British test pilots and the man who test-flew the prototypes, vouched for the aircraft's exceptional and docile handling. Even when flying the early P.1A before a lot of work was done on both it and the second prototype, he wrote that 'far from being a potentially critical and "high-risk" aeroplane, [it] was in fact smooth, undemanding and delightful to fly'.[1]

But that was a highly experienced chief test pilot speaking, not to mention a veteran of Second World War combat. Ordinary mortals doing the conversion course learned that the Lightning was quite a handful in all sorts of ways, no matter that it was the most manoeuvrable Mach 2 all-weather interceptor of its generation. The main problem was always fuel. The Lightning had notoriously short legs, and the pilot who went 'wet' (with full reheat) to 40,000 ft in a couple of minutes from brakes-off found himself with only about fifteen minutes' worth of carefully eked-out fuel left. Never mind looking for Soviet bombers: his priority now was to find either a runway or a Victor tanker, whichever came first. ('Once or twice I swear it was adrenalin rather than AVTUR that kept that thing in the air.') As 'Robby' Robinson once remarked, no ex-Lightning pilot ever ran out of petrol in a car since glancing at the fuel gauges every three seconds became pure habit. It was really the type's chief defect, and ultimately dealt a fatal blow to its wider export possibilities. You could just about push it to Mach 2.3, but only for a couple of minutes, whereas the Mirage III could maintain Mach 2 for twenty minutes. A glance at the Lightning's layout immediately suggested why. The fuselage was packed with two large engines and the wings were thin: the space for tankage was thus severely limited so an ungainly 300-gallon ventral tank was added early on. Because there was no room for them in the fuselage, the main wheels had to be made extremely narrow to fold outwards and fit flush into the wings. This reduced the space for fuel still further, and the Lightning even had fuel tanks in its flaps and leading edges. One of the consequences of the ultra-narrow wheels was that landing this fighter at 186 mph at night on a wet runway in a cross-wind could be a real white-knuckle affair. Surviving pilots would sometimes stagger speechless and trembling into the crew room having finished up in a spectacular and embarrassing ground loop. The tall fuselage and angular fin acted together like a sail and the skinny tyres would scrabble for a foothold. They could even slide when taxiing in the wet over the piano keys – the painted black and white stripes at the end of the runway. The trick in a cross-wind squall was to touch down on the upwind edge of the runway and hope that by the time you had rolled out you wouldn't be making great ruts in the grass the other side (as the rain hissed against your red-hot Dunlop Maxaret brakes).

Another consequence of the design's restricted space was that the nose wheel had to retract forwards. The problem with this was that if the pilot didn't get his landing gear up almost as soon as his wheels left the ground and before he was going faster than 220 knots (a speed quickly reached in a Lightning under take-off power), the increasing airflow against the nose wheel held it back so it couldn't retract. It was not uncommon to see a Lightning with the nose wheel still down go into a steep climb as the pilot tried to lose some speed in order to get his wheel up.

But what the hell – these were the sort of minor quirks of the type that everyone learned to live with, like the Lightning's ingrained habit of leaking fuel which now and then caused engine fires. They were all part of flying an aircraft that was on the very edge of the technology's feasibility: a massively complex array of advanced systems shoehorned into a polished, all-metal-finish aircraft that epitomised sheer power and speed. Never before had the RAF had a fighter for which the engineering ground crew were so crucial to keeping it operational. At first there was resistance and even demoralisation among those who had to deal with the aircraft's ferocious mechanical complexities. Its systems were so advanced that its reliability was seldom much better than poor. There was so much to go wrong; and even a minor electrical relay that was sticking was usually buried inaccessibly within a mass of engineering that had to be dismantled before it could be reached. In the worst cases the ratio reached a thousand hours of maintenance for every hour in the air.

Yet in time, as their own skills expanded and became a matter of pride, the crews became converted to this difficult masterpiece of British aeronautical design. During exercises these Line engineers would often work at night beneath arc lamps out on the freezing pan, doing QTR (quick turn-round) checks while the jockey remained strapped up there in his nice warm office, waiting for the next sortie. They would change his still-scorching wheels and the brake parachute in the tail and check the fatigue meter and climb up onto the aircraft's slippery spine to pour Avpin fluid into the little tank up there so the engines would start. The best Lineys knew more about the internal workings of the aircraft than even the pilots themselves, who had eaten and slept Pilot's Notes for weeks before they ever left the ground in the Lightning's two-seater trainer version. Lineys often sat in the cockpit themselves, testing the engines on 'ground runs' when the aircraft was tethered to rings set in the concrete pan and the twin Avons blasted away with a din that was surely audible clear across the galaxy and the Lightning nodded down on its nose wheel oleo under the thrust, shuddering. Some of these engineers could do everything with the aircraft except actually fly it; and even this was put to the test in a famous incident at RAF Lyneham in July 1966.

Lightning XM135 was suffering from a mysterious electrical fault that only showed up when it was accelerating for take-off. In order to see if he could replicate this the engineer commanding the maintenance unit, Wing Commander Walter 'Taffy' Holden, was doing some taxiing tests along Lyneham's spare runway. He was wearing overalls without a helmet; the cockpit canopy was off; all the ejector seat safety pins were in so it couldn't be accidentally triggered; and there were ground locks on the undercarriage to prevent a Liney pulling the wrong knob by mistake and dumping the aircraft on its belly. After several trips up and down the spare runway and finding that the fault still wasn't showing up, the wing commander gave the throttles a bit of a shove to increase speed and inadvertently pushed them through the reheat detent: the spring-loaded gate at maximum ('military') power beyond which the afterburners engage. The Lightning promptly leapt forward with its famous acceleration on full reheat. Near-frozen in panic, Holden fumbled desperately at the throttles with his left hand but couldn't disengage the reheat. His attention was fixed on looming disaster. Not far ahead a petrol bowser was ambling across his runway, and beyond that a big transport aircraft was beginning its take-off run along the main runway directly across his path. By some extraordinary miracle he missed both but once he had crossed the main runway he realised he had only a few hundred yards left before he crashed through the perimeter fence and exploded in a housing estate. Seeing he had reached flying speed, Holden decided he had no option but to go for broke and take off.

So there he was, a ground engineer suddenly finding himself in the air without a cockpit canopy or flying helmet in one of the RAF's crack interceptors, with the undercarriage locked down and the ejector seat inoperable, streaking away from Lyneham in a vaguely northerly direction. It was fortunate that, like other engineering officers of the period, Holden had actually had several hours' flying training on light propeller-driven Chipmunks and Harvards – what a true Lightning pilot would have disdainfully called 'puddle-jumpers'. At this point training took over. He now remembered how to lift the throttles back through the reheat gates and gingerly tried a turn to get the feel of the controls as the speed fell off. Evidently Roland Beamont hadn't lied and the Lightning was indeed flyable even by someone more used to a Tiger Moth.

Holden managed to bring the thing around the circuit the wrong way and tried to line up for a landing but his height and speed were wrong. He went round again and failed a second time. On the third attempt his approach was good. He flared and put XM135 down, knowing to deploy the braking parachute as soon as his wheels hit the runway. His one mistake was automatically to adopt the three-point landing attitude of a tail-dragger so that the

Lightning's tail bumper banged down hard on the concrete and cut the cable of the parachute which promptly collapsed and fell away. Nevertheless, the wing commander managed to brake the Lightning to a standstill a hundred yards from the end of the runway and shut down the Avons before collapsing from reaction.

It was an amazing feat, and could only have been achieved by someone who had not only mastered the rudiments of flying but who knew the Lightning's cockpit from an engineer's point of view. It is said that the poor man's nerves took years to recover. As for XM135, it is on display in the Museum at Duxford, and anyone paying a visit should view it in the respectful light of this epic flight it made one July day all those years ago.

An exceptional, one-off case, maybe; but the best ground crews became so attuned to the job that there could grow up a sort of three-way symbiotic relationship between pilot, Lightning and the men on the ground. A typical exercise might indeed take place on a cold winter's night with rain falling. On this sort of quick turn-round caper the Lineys' job is to pit-stop the Lightnings and get them back into the air without delay. Because there is so much hazard attached to the task, especially at night, the wheel-changing and QTR checks are carried out with nearly balletic precision. Once the checks are complete the crew clear the area around the aircraft. This has become cluttered with nitrogen tanks for the tyres, wheel jacks, sawn-off drums placed strategically around beneath fuel leaks and a sack of what looks like cat litter (known to all as 'chicken shit') used to mop up a spill of OM-15 hydraulic fluid. Soon she's ready to go. The Lineys can follow the pilot's cockpit rituals of start-up as though they themselves were sitting up there beneath the rain-speckled canopy. Main battery ON. The big Houchin generator down on the pan is already plugged into the aircraft and chugging away, feeding in external AC and DC voltage. Pilot now checks that both throttles are moving freely before giving the cautionary wave to show he's ready to start no. 1 engine. This is when you make quite sure there's nobody still lurking in the shadows beneath the Lightning because he could get horribly burnt. In order to start a jet engine it is necessary to get it turning over fast enough and hot enough so that when fuel is fed in it will ignite. Some aircraft use starter cartridges filled with cordite: a muffled explosion and clouds of smoke puffing out of the starter exhaust so that – in the case of the Javelin, for instance – the aircraft will be squatting there in a heavy black cloud, its undercarriage temporarily invisible like the feet of a sixties pop group standing onstage in billows of dry ice. The Lightning, like the Hunter, doesn't use cartridges to get its Avons going: it uses Avpin. This is the brand name of isopropyl nitrate (or IPN): a highly volatile liquid fed from a small tank up on the Lightning's spine. When it ignites in the starter, blazing gases vent under the aircraft and it is a serious error to be standing anywhere nearby. Apart from the danger of burns, once ignited Avpin gives off characteristic fumes that act like tear gas on anyone unwise enough to get too close.

So you signal up to the pilot with a circular 'wind-up' gesture that the crew are clear and he opens the throttle of no. 1 engine (the bottom one) which in turn opens the high-pressure fuel cock. He pulls up the rapid start gang bar – low-pressure cocks ON – flicks the start switch to START and presses the start button on the console by his right knee. There is an immediate hissing scream as a couple of litres of Avpin ignite. This is one of the great familiar sounds of a Lightning unit, as much a background to life on the station as a cock's crow is on a farm. It lasts several seconds as it spools the heavy engine up to speed. In the shadow of the aircraft's belly a faint blue St Elmo's fire of dancing flame flutters around the starter vent. Everyone has carefully avoided standing in line with the port wing. Occasionally, the exploding Avpin shatters the blades of the starter and they can fly out in this direction at bullet-like velocity. This time all is well; the hissing fades beneath the Avon's rising roar as it begins to run on its normal diet of AVTUR and no. 1 is safely lit.

Since it's a moderately foul night and Lightnings can be temperamental when starting in the wet the pilot has not signalled to disconnect the external power even though he can generally start no. 2 engine by using the power from no. 1. More of the shrieking Avpin hiss and eye-watering fumes until the second engine catches. Up there in the cockpit the pilot's eyes will be glued to the rpm and jet-pipe temperature gauges which move swiftly. Even on a cold night like this the engines will reach 800° centigrade in a matter of five seconds. When everything looks stabilised and normal and the intake pipe is roaring with gulped air only inches beneath his backside he's pretty much ready to roll. The crew disconnects the heavy generator and wheels it off to one side. All the aircraft's nav lights are on and the pilot can be seen checking his canopy is shut and the ejector seat pins are out and stowed. A glance down to his left to check the brake-pressure gauge and he signals he's ready to move with a quick flash of the taxi light. You signal back that all chocks are clear. He flicks off the bicycle-style brake lever on the control stick and begins rolling.

Now is the time for a wise ground crew to shelter behind the Houchin generator to avoid a roasting as the Lightning turns on the pan. Even throttled back to taxiing revs the Avons are shatteringly loud. The twin tailpipes swing around like over-and-under gun muzzles and blare furnace heat. Anyone caught in the open hunches and turns his back, the searing blast flattening the creases in his weatherproof gear and setting the protruding cuffs of his kerosene-proof overalls flapping. Only a minute or two later their charge is reduced to a distant clutch of lights speeding along the runway, followed by what seem to be twin blowlamps in whose fiery cones strings of shock diamonds dance. Then they tilt and the Lightning quickly vanishes, towing its thunder up into the overcast. Their pilot is gone, leaving behind him the muscular fragrance of burnt kerosene and Avpin fumes.

Somewhere out over the North Sea he will rendezvous with a Victor tanker and attempt the tricky task of night refuelling. This entails coming in behind the tanker while carefully avoiding the larger aircraft's wing-tip vortices which can flip an unwary Lightning over like a dead oak leaf. Then it's a matter of matching the tanker's speed and getting the probe that sticks out just to the left of the cockpit to nuzzle up into the trailing basket (like trying to push spaghetti up a bear's backside, as someone memorably described it). But up there in the moonlight above the clouds, tiptoeing along with a gigantic V-bomber filling his windscreen as he nudges the basket hard enough to open the valve and let the AVTUR gush into his tanks, he's in pilot heaven. As far as he's concerned he's flying the world's greatest fighter and what he's doing is the apogee of living, his alertness knife-edged as he makes tiny corrections to the controls, eyes flicking back and forth between the view ahead and the instrument display as he burbles in the tanker's wake before breaking away with cold moonlight flashing off the wings of his metal bird. He wouldn't swap his Martin-Baker seat with anybody, not for a king's ransom.

The Lightning was designed purely as an interceptor: to get to height as quickly as possible after take-off in order to challenge – and if necessary destroy – incoming Soviet bombers. Its manoeuvrability at altitude was also intentional in case it had to dog-fight an equivalent Soviet fighter. After the engagement it would either home in on a tanker or recover to base, tanks nearly dry and tailpipes glowing. Its primary need was thus for sheer performance in terms of climb rate, top speed and altitude, a superiority that was absolute when it entered service. In other respects, though, the Lightning was not quite so dazzling. Not many years into its service its armament, as well as its radar and avionics generally, became markedly inferior to those of contemporary foreign interceptors. (Wing Commander R. S. Hargreaves, describing firing two-inch ballistic air-to-air rockets from his Lightning in the mid-sixties, said that 'sitting behind these amazingly inaccurate devices was a bit like seeing an open box of matches thrown in front of you, each match with a mind of its own but no idea where it was supposed to be going'.[2]) The main reason its weapons and radar systems were never properly updated was because the Lightning was planned for only ten years' service and it didn't seem worthwhile. In actual fact it was to last twenty-eight years with the RAF, by which time its weapons and avionics were frankly obsolete. Its performance, however, remained supreme. In particular the earlier marks, before 'weight creep' began to eat into pure performance, could still have shown most of today's jet fighters a clean pair of heels – and that by an aircraft designed nearly sixty years ago. It is worth noting in passing how radically different from each other were the Lightning and its contemporaries, the F-104 Starfighter and Dassault's Mirage III, even though all three designs were intended to do pretty much the same job. It was a testament to the still imperfect knowledge of supersonic aerodynamics in the early fifties that each of so many different solutions (sweepback, stub wings, delta) seemed so full of promise. Well over fifty years on, and with almost everything now understood about the behaviour of practically any aerofoil up to some 7,000 mph, it can be seen that designs for the same role have tended to converge, just as Boeings and Airbuses look more or less indistinguishable.

Lightning devotees like to argue interminably about how its phenomenal performance compared with that of its contemporaries. Its initial climb rate was 50,000 ft per minute, whereas the Mirage IIIE could manage only 30,000. In his F.Mk 3 Lightning Flight Lieutenant Mike

Hale took part in time-to-height and acceleration trials against F-104 Starfighters and reported that the Lightning won all the races easily except that for low-level supersonic acceleration, which was a dead heat. Hale also gave an American U-2 'spy plane' a nasty shock by intercepting it in his Lightning at 88,000 ft, an altitude where U-2 pilots had hitherto considered themselves safe from molestation. Years afterwards the late Brian Carroll, former Lightning pilot and ex-Lightning chief examiner, exchange-flew the McDonnell Douglas F-15 Eagle, which was developed some eighteen years after the British aircraft. He found the Lightning was still quicker off the ground than the Eagle. He concluded: 'Both aircraft had very similar performance and handling characteristics, both were a joy to fly ... Given the choice between the two, I would have been mad not to take the Eagle, but only because it has such superb avionics and weaponry. However, for the pure joy of flying, the Lightning still heads the list ... A magnificent aircraft.'[3]

Magnificent, undoubtedly; but in true British fashion the Lightning's sales never reflected its outstanding abilities. In all, only 337 were built (in comparison with 2,221 Starfighters and 1,422 Mirage IIIs). After the type's twenty-eight years' service, the last RAF Lightning was grounded in 1988. Leaving aside the special cases of the Harrier and the Anglo-French Concorde, the Lightning could be said to represent the summation of British aeronautical design. The country produced more aesthetic-looking aircraft (the Hawker Hunter, for one) but nothing entirely home-grown that was so technologically advanced for its day as well as so fabulous to fly. It was an aircraft quite unlike any other. Merely seeing it put through its paces could afford a middle-aged New Elizabethan spectator an experience that felt borderline transcendental. At a Biggin Hill air display some time in the early seventies I watched a formation of nine Lightnings perform as one. Their precision was perfect: I never saw a better mass aerobatic display anywhere, before or since. To glimpse them scudding low over the distant boundary like some wheeling prehistoric flock, banking around in the haze to line up with the runway and approach along it at fifty feet in near silence was to feel the hair on your neck bristle in anticipation. Then suddenly here they were, closing in with white vapour shimmering over their wings before all nine of them reared up on pillars of smoke directly in front of us and blasted skywards like a single sheet of metal bent. The battering of eighteen Avons on reheat seemed enough to jar the planet from its orbit. We were cocooned in thunder, annihilated by din. Grown men with idiotic fixed grins on their faces, half-deaf, turned compulsively to perfect strangers shouting, 'Did you *see* that? *Jesus! Woooh!*' while far overhead nine silver scratches on the sky's surface re-formed into an arrowhead poised to plunge once more and transfix the green heart of Kent. It was beyond exhilarating. It eclipsed that conventional fifties pride and patriotism we used to carry with us to air shows like a packed lunch. This was three hundred combined tons of thrust under fingertip control becoming a single entity: a brutal, beautiful efflorescence of pure technology and consummate skill. These days, if I feel my eyes still misting slightly at the memory, it must be partly from regret that I shall never see such a thing again, and nor will anybody else. None of today's air show-goers will ever see nine Lightnings flying as one. The usual abject combination of penury and dreary safety regulations means that only if they're very lucky will most Britons see a lone example of the flower of their country's aviation genius trundling tamely up and down a runway in front of them. Watching a Lightning taxi is better than nothing, I suppose. An ex-Lightning pilot, now teetering on the edge of his eighties, has words for a luckless modern generation. 'Next time you watch one of those motoring programmes on TV, remember that to someone who has flown a Lightning, driving a Ferrari is about as exciting as driving a Trabant.'

As it happened, but for politics the Lightning might not have gone unchallenged as the flower of British aviation genius. Even as they were working on its prototype English Electric were pondering a successor to their Canberra. This had been enormously successful and was clearly destined for many years' service yet in air forces around the world; but such was the pace of aerodynamic progress that by the mid-fifties it was fast becoming technologically obsolete. Even before the Air Staff had issued their General Operational Requirement (GOR 339) for its replacement, English Electric had sketched out an aircraft whose performance would be more ambitious even than that of their new Lightning. GOR 339 was the typical wish-list of air force zealots everywhere in that it called for an aircraft that would do almost anything except actually travel under water. The RAF wanted something that could fly in all weathers at supersonic speed at both high and low altitude;

carry a large payload of tactical nuclear weapons; have advanced avionics; have equally advanced reconnaissance capability (like the Canberra's but far superior); be able to operate from short and even improvised strips; and finally have a long range. One could almost imagine these air chiefs sitting around watching an episode of *Thunderbirds* and exclaiming, 'Yes! We'll have one of those!', except that that series still lay ten years in the future.

The design that emerged from these near-impossible specifications was TSR.2 (the initials standing for Tactical Strike Reconnaissance). As we have seen, by the time Sandys published his White Paper, English Electric's project, like its Lightning, was far enough advanced to escape his axe; but nearly two more years went by before the government got round to awarding a contract for it, which it did on 1 January 1959: a delay that was evidence of indecision at the top. By then the provisions of Sandys's policy had been implemented and English Electric, like practically all other British aircraft firms, no longer existed as an independent company. It had become part of the group that included Vickers-Armstrong and Bristol Siddeley, together forming the nucleus of what was to become the British Aircraft Corporation in 1960. The government's TSR.2 contract designated Vickers-Armstrong as the lead firm, while Bristol Siddeley Engines were asked to supply the aircraft's power plant. Shocked, English Electric found themselves cheated out of building their own aircraft. They were relegated to being sub-contractors but with the sop of a guaranteed half-share of all the airframe work. Given their experience with the Lightning – unique in the industry, since nobody else had built a Mach 2 aircraft – it was hardly surprising that this decree left the company feeling badly miffed. It was not an auspicious way to manage the most advanced aircraft project that Britain had yet conceived.

Many of the seeds of the TSR.2 disaster lay buried in this arrangement, but there was a kind of reasoning behind the apparent madness. There were some very powerful egos caught up in the project, and none of the old grandees of British aviation had liked this policy of forced amalgamation. By now 'Teddy' Petter had left English Electric to go to Folland and his place as chief designer had been taken by the equally brilliant Frederick Page, who had been largely responsible for taking over Petter's Lightning and developing it further. TSR.2 was a Page design, but unfortunately he and his team of draughtsmen at Warton in Lancashire were barely on speaking terms with their own firm's autocratic Arthur Sheffield, who still headed English Electric's old production team at Preston. This situation did not inspire confidence; and for this reason Vickers-Armstrong (firmly under the sole control of the charismatic George Edwards) were awarded the lead contract. To government officials this no doubt seemed a pragmatic solution but it was scarcely a recipe for smooth co-operation on an aircraft that was far more complex and advanced even than the Lightning. Consequently, internal relations within the group were frequently dreadful, which only added still more months of delay to TSR.2's already slow progress. Nor were external relations any better, given the constant sniping from the Treasury over rising costs, from the Royal Navy in the person of Lord Mountbatten, and from the MoD's chief scientific adviser, Sir Solly Zuckerman. The advent of Harold Wilson's cost-cutting Labour government in 1964 merely put the cap on it.

The reason the navy were against TSR.2 was that they had already been promised a carrier-borne strike aircraft for their own use, the Blackburn Buccaneer, and they worried that this fancy project of the RAF's would mop up all available funds before their Buccaneer could get airborne. Seeing how TSR.2's costs were increasing at a near-exponential rate, the Treasury and Westminster were by now hoping the RAF might abandon the semi-stalled project and sensibly opt for its own version of the Buccaneer. Things would be so much simpler... Unfortunately, the Buccaneer was not even going to be supersonic, and the RAF had set its mind on an aircraft with at least Mach 2 performance. After the Lightning, anything less potent would seem like a humiliating step backwards. How could they possibly hold their heads up among the Americans and the French and the Russians if they were once again to be relegated to the subsonic league? Additional heat was supplied to this argument by the ancient inter-service rivalry in which the RAF and the Fleet Air Arm each pretended the other hardly existed, let alone was competent to fly an aeroplane. Blackburn accommodatingly came up with a proposal for the P.150, a supersonic version of the Buccaneer that would have fulfilled most of GOR 339's requirements; but the RAF's continued lack of interest was studied to the point of being a snub.

Such background details are inseparable from the TSR.2 story. Each hampering strand – whether of internal bad blood or of external political opposition – acted like the Lilliputians' ropes that tied down Gulliver. And yet the miracle is that four prototypes of TSR.2 were built, and one even flew, and reportedly flew well. The resulting aircraft was a big two-seater with a fully laden take-off weight of over forty-three tons. It was perhaps impressive-looking rather than beautiful. The design was clearly influenced by a foreign contemporary, North American's carrier-borne A-5 Vigilante bomber, although the British design was more laterally compressed into a narrower, slab-sided short delta whose wing tips were bent downwards in a brooding anhedral. Behind and below this wing was a smaller deltoid tailplane with a single swept-back fin. To make up for its skimpy wing TSR.2 had fully blown flaps, thrust being bled from the engines and directed out of holes on the flying surfaces to augment the effectiveness of the pilot's controls. The wings were high mounted with the intakes tucked, one on each side, into the fuselage beneath them. Ahead of them was a long beak-like nose with the tandem cockpit well forward.

Once again it fell to Roland Beamont to test fly the prototype. After English Electric was swallowed up he had been given the choice of becoming BAC's chief test pilot to concentrate on testing Concorde, because after his vast number of hours on Lightnings he had more Mach 2 experience than almost anybody else. In the event he turned the offer down in favour of flying TSR.2 and Brian Trubshaw got the BAC job. So Beamont took TSR.2 south to Boscombe Down, which had a longer runway than Warton, and began taxiing trials in 1964.

The aircraft had taken almost a decade to reach this point. This was supposed to be the strike/reconnaissance type on which the whole of the RAF's future was to be based from the late sixties and into the nineties and beyond. Beamont himself later described its genesis as 'a classic example of how not to procure a new aircraft'. Yet out of this familiar blend of Whitehall farce and industrial confusion a large white-painted aircraft, XR219, duly performed its taxiing trials at Boscombe and on 27 September finally took to the air. In so doing, it represented a huge leap ahead of any other European military aircraft. It was designed to fly supersonically as low as 200 ft using ground-hugging radar, and both cockpits had very advanced head-up displays (HUD) which projected instrument readings directly on to the windscreen. If the avionics were state-of-the-art, the aircraft's Bristol Olympus engines were still rather less so; they had long been giving severe problems and had been approved for only this single outing. This merely placed the occasion well within the traditional 'by guess and by God' approach of a British maiden flight. Two smaller aircraft were designated to fly 'chase', accompanying XR219 in the air to take pictures and to relay any information about his aircraft that Beamont couldn't see for himself. One of these chase machines was an all-metal-finish Lightning, flown by Jimmy Dell, by then Beamont's deputy at Warton. The other was a bright yellow two-seat Meteor NF.Mk 14 with a clear bubble canopy for photography.

One of the pilots nominated for the Meteor was 'Robby' Robinson, but to his disappointment it turned out that on the day it was his colleague's turn to take the official photographer up. He watched Beamont take off with fast acceleration and a steep climb-out with reheat, twin chains of diamond shock waves clearly visible in the fiery exhausts. Then Robinson drove over to the air-traffic control tower and eavesdropped on the radio chatter between Beamont in TSR.2 and Dell in the T.Mk 4 Lightning. Eventually the TSR.2 reappeared in the distance flanked by the Lightning to its port and the Meteor to starboard, a colourful trio led by the great white hope of British aviation. Once landed, Beamont and his observer, Don Bowen, said how impressed they were by the new aircraft's handling, despite the limits on the engines. As Beamont reported,

From take-off onwards throughout the flight I could not fault the control and stability ... Three-axis responses had been so good [that] this untried prototype might just as well have been an already developed and proved aeroplane ... This precise controllability was apparent right down to the landing. The flare to a low-rate touchdown was so straightforward that the next thing was a sharp surprise. As the main wheels touched, a violent oscillation occurred at the cockpit, so severe that it momentarily blurred my vision. But this cleared as I throttled back and the speed dropped and then, still nose-high at about 140 KIAS [Knots Indicated Air Speed], I streamed the drag 'chute and this fascinating first flight came to an end.[4]

Following this maiden flight Beamont wrote in his report: 'In this configuration and under these test conditions TSR.2 could be flown safely by any pilot qualified on Lightnings or similar aircraft.' It was an astonishing testimonial for a first flight, as was Jimmy

Dell's after his own first flight some time later: 'It's a pilot's aeroplane all the way.' The tests continued; the landing gear problems were gradually solved. They flew the aircraft to 500 KIAS very low, at supersonic speed at 30,000 ft, and at 500 KIAS through tight Pennine valleys within fifty feet of the hillsides, all with the same high degree of controllability. TSR.2 was fast shaping up as a most exceptional aircraft: certainly a worthy successor to the Lightning. The test schedule was interrupted for a couple of months during which new engines were readied and installed. This was immediately seized on by the anti-TSR.2 lobby as proving that the aircraft was no more than a bottomless hole down which yet more money was being recklessly poured. The RAF (they declared) should now cut its losses and buy the American General Dynamics F-111A swing-wing bomber instead. Then suddenly on 6 April 1965 the Chancellor, James Callaghan, announced that TSR.2 was to be cancelled immediately. The news reached the test pilots at Boscombe as they were having lunch in a local pub. They immediately tore back to the airfield to get the second prototype into the air in the hopes of showing the government that the project was far too advanced to terminate, but were refused permission to take off.

Thus ended TSR.2, and with it all hopes that Britain could remain in the major league of aircraft manufacturing countries. £195 million in development costs were written off. The immediate effect on BAC, and particularly on English Electric's Warton workforce, was devastating, with large numbers of redundancies and a general demoralisation that hit the RAF and spread throughout the industry. Fairey's Rotodyne feeder helicopter for ferrying passengers from airports to city centres had already been cancelled (£21 million written off). Now, axed together with TSR.2 were Hawker's P.1154 projected supersonic version of what was to become the Harrier vertical take-off fighter (another £21 million) and the Hawker Siddeley/Armstrong Whitworth 681 military freighter (much the same amount). There was widespread disbelief at both the policy and the crassness with which it was implemented, but this quickly turned to real anger when the full irony was revealed: that the 150 F-111As ordered for the RAF because they would have been so much cheaper than 150 TSR.2s turned out – when equivalent F-111As were delivered to the Australian Air Force – to be even more expensive because their recurrent bugs and teething problems cost a fortune to put right. Eventually the UK's order for F-111As was itself cancelled in January 1968 (at a cost of £46.4 million). This left the RAF without any long-range strike and reconnaissance capability. Yet more irony was to follow, because the RAF were then obliged to accept the navy's Buccaneer as a stop gap and – amazing to relate – they even learned to love it a bit. It was a fine aircraft of its kind, in a formidably strong but old-fashioned, subsonic sort of way, and it gave the Fleet Air Arm in particular good service. Its low-level attack capability was probably unequalled by any other fighter anywhere, but no hot jock ever for a moment thought the Bucc was a substitute for a Lightning.

As was the usual practice, the order went out for the immediate destruction of the TSR.2 prototypes and all the jigs. BAC management at Warton entered a desperate final plea that a single prototype should be spared for high-speed tests associated with Concorde, but to no avail. In 1966, during a visit to Warton, the aviation journalist Don Middleton asked a senior BAC official why the government hadn't allowed them to use a TSR.2 in the Concorde programme. 'His answer was, "Wouldn't it have been embarrassing for the Government when the tests revealed the true measure of TSR.2's superiority over the F-111A!"'[5] With the sole exception of the Harrier, Britain was never again to produce an unequivocal world-class aircraft on its own. How the cancellation of its much-needed new strike aircraft as well as its new transport were the answer to the RAF's being 'dangerously overstretched and seriously under-equipped' (in the words of the Labour government) is no clearer now than it was then.

The summary ending of TSR.2 just when it was showing exceptional promise is a subject that still arouses passionate opinions over half a century after the event. Any engineers or pilots who were involved in the project not unnaturally tend towards incandescence in their memoirs. Sir Stanley Hooker, at what had been Bristol's Engine Division, wrote of how they were at last getting the better of the problems in TSR.2's Olympus engine. (That they did was borne out by its later development's extremely reliable use in Concorde.) He added: 'This wholesale bout of cancellations brought to an end ten years of the worst mismanagement of the RAF's equipment and of the British aircraft industry that could possibly have been arranged.'[6] It is perhaps fortunate that Hooker, who died in 1984, never lived to see what the Ministry of Defence and various defence companies between them could *really* achieve by way of mismanaging the RAF's equipment – not to mention that of the army – a full quarter-century after his death.* As for the celebrated Fleet Air Arm pilot, Eric 'Winkle' Brown, TSR.2's death sentence filled him with justified foreboding. 'What appalled me was the vicious follow-up to the cancellation, for all the production jigs were destroyed and the prototype aircraft scrapped or sent to museums. To my mind this was a portent of what might lie ahead for the Services . . .'7

As TSR.2's chief test pilot, Roland Beamont was characteristically outspoken. 'The widespread smear campaign of 1964/65 implying that the TSR.2 was a scandalous technical and industrial failure and a waste of taxpayers' money can be seen for what it was, as a monstrous manoeuvre for political ends from which the striking power of the Royal Air Force and the world-leader potential of British military aviation technology have never fully recovered . . .'8

Because the argument is so polarised it is only fair – as well as interesting – to present an opposing case and point out that some very knowledgeable professional aviators have strongly defended the fateful cancellations of 1965. One of these is the eminent test pilot John Farley, who

---

* In 2006 RAF Nimrod XV230 suffered a mid-air fuel leak and fire over Afghanistan, the subsequent crash killing fourteen men. A report in 2009 named and shamed five people from the MoD, three from BAE Systems and two from QinetiQ as effectively responsible for the disaster by having ruthlessly pursued cost-cutting at the expense of safety. Also in 2009, a leaked official report by Bernard Gray on the MoD's procurement methods gave a withering assessment that caused acute political embarrassment at a time when the UK press was full of complaints by British military figures in Afghanistan about shortages of vital supplies such as helicopters and armoured personnel carriers.

became internationally celebrated for his testing of the Harrier and for demonstrating it at sundry air shows in the seventies and eighties in some of the most accomplished flying displays ever seen. In a conversation in early 2009 Farley made a case for not considering the British aviation industry as having been a mere helpless victim tragically wronged by calculating politicians. One could almost hear the shades of Bill Waterton – whose own sceptical opinion of the industry he had worked for was no less pungent – cheering him on. Farley began by describing the way the system worked at the time:

It operated on a Cost-Plus basis. The idea was that we [in the industry] spend money, we do research, we do development, we give you [the taxpayer] the receipts, you pay the bill – and then you give us a percentage on top as profit . . . This was a *ludicrous* way to run an aircraft industry. But it was how those four projects were organised. And in order to get the spending back under control the Labour Government, in my view quite correctly, cancelled the lot. Now, I don't believe they cancelled them for the right reasons – for the technical reasons. I think they cancelled them from a political point of view in order to get the industry under control and to get some more sensible contracts organised, as well as to save money.

The technical reasons? Take TSR.2. The spec. was for a very high supersonic, low-level, long-range delivery of a nuclear weapon: a plane that's got to fly very fast, very low, and a very long way. The only way to do that is to put a negligible wing on it so you don't get any drag, and fill it up with fuel. If you like, a bit like the Starfighter which was designed just to have a high top speed but negligible wing. The trouble with a negligible wing is your take-off and landing performance becomes appalling. And if you look at the films of the one TSR.2 that flew – and that was flying at light flight-test weights, not operational weights – you see the large angle of attack needed and the long ground run and the very poor landing performance. Even if they had accommodated that, the aircraft was still going to have very poor manoeuvring capability. If we'd had the TSR.2 I think some would still be in service today with the Royal Air Force. And in my view, being caught out in the open by MiGs or Sukhois or ground-to-air missiles, you would have had no manoeuvrability whatsoever with that little wing. I've never flown an aircraft yet when I didn't feel a bit short of wing – except the Vulcan.

So I didn't like the TSR.2 because it didn't have enough wing. And I didn't like the P.1154 [the projected supersonic VSTOL (Vertical/Short Take-Off/Landing) antecedent of the Harrier] because as a pilot I thought the concept was stupid in the extreme. The whole point of a VSTOL aircraft is its operating flexibility: it can fly from a field, a bit of road, the back of a ship, anything. But if you want that, you've got to moderate your exhaust gas velocity and temperature which both need to be very high if you want supersonic performance. If you don't moderate them you're not only going to set fire to the landscape and melt the landing surfaces, including probably your own tyres, but when the jets are horizontal you're going to damage the fuselage and tail unless you use some pretty exotic materials because you're washing the whole side of the fuselage in what is the equivalent of a re-heat system. No, we knew at Bedford in 1966 that the P.1154 was a stupid bloody idea.

So what about the A.W.681 [the projected military transport]? It was really a hovering mini-Hercules. The idea was that you would land this large transport aeroplane in some farmer's field and deliver – for example – a new engine to a P.1154 that needed one and was standing in a corner of the field. Well, how ridiculous to design and develop a transport aeroplane to do that when what's wrong with a helicopter? It was a non-starter. And as for the Fairey Rotodyne, they couldn't build a gearbox strong enough to handle the necessary torque so they opted for tip-jets to drive the rotor. Did you ever see or hear a Fairey Rotodyne fly? From two miles it would stop a conversation. I mean, the noise of those little jets on the tips of the rotor was just indescribable. So what have we got? The noisiest hovering vehicle the world has yet come up with and you're going to stick it in the middle of a city . . . ?[9]

John Farley's arguments sound cogent, especially those concerning TSR.2. He clearly feels that no amount of craftily blown flaps can make up for an absence of wing, and the shape of most modern Russian, French and American fighters today suggests he is right. After all, he is speaking both as a pilot and as an engineer. But then so was Roland Beamont a pilot, and he seems to have been not at all dismayed by the aircraft's 'negligible' wing. However, another ex-test pilot (who prefers to remain anonymous) maybe shone some light on this when he told me he always believed Beamont had blundered by electing to test TSR.2 instead of Concorde. He said Beamont – like most others in the profession – had never seriously believed the military aircraft could be cancelled outright, partly because the project was too far advanced and partly because at that time there was a lot of well-informed gossip to the effect that the British government secretly wanted to pull out of the Concorde project. So perhaps he reasoned that hitching his fortunes to the Anglo-French aircraft might be unwise and was content to leave it to Brian Trubshaw. (The UK government did indeed try to renege on the Concorde agreement and were only dissuaded by being told how much more it would cost them if and when the French quite rightly exacted punitive damages.) My informant's point here is that having belatedly discovered that he'd backed the wrong horse, Beamont had no face-saving option but to claim TSR.2 was a far better aircraft than it actually was. To some extent he was also identifying the waning of his own career with its demise, thereby according its cancellation the status of grand tragedy... Well, it's a theory. I shan't speculate further.

In any case, the type with which John Farley is most identified, the Harrier, has the sad distinction of being the very last great aircraft Britain produced unaided. (I don't count BAe's little Hawk trainer despite it having been bought in numbers by the RAF as well as being the third post-war British aircraft to be bought by the US – with over 300 in USN service as the McDonnell Douglas T-45 Goshawk). Even so, it took the Falklands campaign and twenty-two confirmed combat victories to convince people of just what a brilliantly flexible machine the navy's Sea Harrier was. It is significant for being only the second British post-war aircraft (after the Canberra) to be bought by the United States. As with the Canberra, the Americans modified the Harrier for their own use, principally by the Marines. The fact is that for decades there was no other operational aircraft in the world like it although several countries tried to design a Vertical Take-Off/Landing 'jump-jet' that combined the advantages of a helicopter with the abilities of a conventional fighter.

This was a logical enough dream and had quite a history. Many a New Elizabethan's breakfast table in 1953 was enlivened by newspaper photographs of the contraption known as the 'Flying Bedstead'. This was an attempt to get controllable flight from an ungainly rig that consisted of a downwards-facing jet engine, some of whose thrust was bled off through small nozzles at the four corners to maintain stability. It looked nothing like an aircraft, and wasn't intended to. What we didn't realise at the time was that there was bitter rivalry between engineers who favoured using several small engines, and those who were backing the 'Flying Bedstead' model of a single large jet engine with nozzles that the pilot could swivel. Other nations were experimenting with similar – and even more outlandish – options. Both France and the United States came up with experimental designs that sat on their tails and required the pilot to take off and land while lying on his back. Only the British came up with a flying design that so elegantly solved the complex engineering problems yet retained the characteristics of a high-performance strike/attack aircraft. The Harrier's vectored-thrust system (i.e. a single main engine whose thrust could be ducted to various parts of the airframe and directed at various angles) proved far superior to alternatives that incorporated several small engines with fixed thrust, all of which added redundant weight when not in use. So effective was the system that after a Harrier prototype touched down on *Ark Royal* in the first-ever landing of a VTOL aircraft on a carrier at sea, it took another forty years for aerospace engineers to come up with a truly modern version of vectored thrust. Jets like the Sukhoi

Su-47 Berkut and the fifth-generation Lockheed Martin F-35 Joint Strike Fighter (JSF), with their thrust vectoring and computerised fly-by-wire controls, offer a new approach to making aircraft highly manoeuvrable in the air as well as having VSTOL ability. With plenty of wing (the Sukhoi's sweep *forward*), paddle-like tailplanes and splayed fins they are recognisably modern fighter aircraft. It is only when one sees an F-35's jet pipe swivel through an extraordinary range of angles that one realises how such aircraft achieve their short take-offs and vertical landings, not to mention their amazing agility in the air. It is significant that British Aerospace was part of the team that developed the F-35.

With the Harrier, seventy-odd years of independent and often world-leading aeroplane design and manufacture were drawing to a close. From now on, any British aircraft venture of any size would involve collaboration. The expense of designing, building, testing and developing a completely new aircraft was simply too much for the UK to bear unaided. Joint projects, often involving several foreign partners, became the order of the day: the SEPECAT Jaguar, Panavia's Tornado, the Eurofighter and, of course, the granddaddy of them all, Concorde. In terms of public perception Concorde's record as an aesthetic and techno-logical triumph still tends to outweigh its monumental commercial failure. Never mind the noise – it looked fabulous, it was the only thing of its kind in international service, *and it wasn't American*. Somehow, people managed to overlook its massive development costs (around £1.1 billion) and that it was only ever flown by a tiny handful of bankers, politicians and celebs. This meant that France and Britain's roughly thirty-four million families each paid over £33 in taxes to enable people like Mick Jagger to fly from London to New York in three and a half hours instead of eight. Democratic it wasn't. But then, of course, neither would the Brabazon have been.

Until the era of enforced collaboration dawned and national pride had to be swallowed, it should be recorded that, despite having had to labour under managerial, military and political incompetence that would have disgraced a banana republic, Britain's aviation industry could draw on a huge national resource of technical expertise to produce some outstanding aircraft. This expertise, of course, began to be dissipated as soon as much of the design, development and manufacture shifted overseas. But we ought to remember that before this happened the range of our aircraft industry's competence included guided missiles and rockets and even extended into space. In the sixties what had been Saunders-Roe and Westland Aircraft (before they were swallowed up by BAC) produced Black Arrow, a satellite-launching rocket that in October 1971 put Britain's first and only satellite, Prospero, into orbit from Woomera in Australia. Naturally, in keeping with British tradition, the project was immediately cancelled once it was a success. Wikipedia comments: 'As of 2009, the United Kingdom is the only country to have successfully developed and then abandoned a satellite launch capability.' 'And an aircraft industry,' it might have added; 'not to mention just about all other industry, come to that.' But just as relics of a once-great aviation tradition can still be seen in museums and at air shows, so brave little 66-kilogram Prospero is still whizzing around in low earth orbit. Probably few Britons even realise it is up there; but it is likely to outlast them because its orbit will not decay fatally for another century or so. Until then, as the character after whom it was named remarked: 'space enough/ Have I in such a prison.' The momentary streak of incandescent gases across the upper atmosphere that will mark Prospero's passing will be a future generation's reminder of what Britain could once do.

Not with a Bang

In June 2009 I stood beside Bill Waterton's austere and harmonious headstone in the little cemetery of Oxenden, Ontario. I was too late by three years to have met my boyhood hero, which was probably better for both of us. He had wished to be buried by Georgian Bay, and the water was close by. Even closer was the end of the runway of Wiarton's small airport. It was not difficult to imagine the old instructor lying beneath my feet in his coffin, head pillowed on the very flying jacket he had been issued when he had first joined the RAF, critically monitoring the approaches and take-offs of the mainly light aircraft that occasionally flew low above him. For most of the time, though, the place would be one of deep silence.

Silence, too, had attended his passing. The Owen Sound *Sun Times* had interviewed its once-celebrated local resident back in 2003 to celebrate the hundredth anniversary of the Wright brothers' first flight but, at Waterton's own request, printed no notice of his death in 2006. The Toronto *Globe and Mail* did give him an obituary, but not until four months after he had died. Three thousand miles away in England, where virtually all his true professional career and achievements had taken place, there was not a mention of him, not even in the aviation press. It was as if Bill Waterton had never been. Yet this was surely no deliberate snub born of rancour, the afterglow of a fifty-year-old vendetta. The outrage must long since have evaporated that once was felt by a beleaguered aero industry now changed beyond recognition, and by bruised and slighted executives who departed the scene decades ago. Curiously, Waterton himself went on to the end of his life speaking almost with relish of being persona non grata in both Britain and Canada. I wondered whether even as a youngster he might have abandoned hope of getting on easily with people and instead had cultivated a perverse satisfaction in the idea of being universally blackballed, persisting in this conviction long after most people could even remember why. Surely the reason Waterton's name has never been rehabilitated and his legacy re-examined is simply because to all intents and purposes he vanished into thin air back in 1956. He was forty when his book was published and he was sacked by the *Daily Express*, and he was ninety when he died. In that intervening half-century he became invisible to his former world: to ex-colleagues and his readership alike. From their point of view it was as if his life had ended in 1956. For pure want of revision, their image of him as aviation's black sheep gradually fossilised before crumbling into forgetfulness, even as it lingered on here and there in half-remembered tales in both Britain and Canada.

After leaving the *Express* he, Marjorie and their young son Willy spent three gipsy years wandering around southern Europe and North Africa in a Volkswagen van ('the smartest thing we ever did'). It was a decade before such a thing became fashionably hippie and it seems likely that Waterton, always curious, had acquired a taste for exotic travel during his trips to Turkey and the Middle East while demonstrating and selling Meteors for Gloster. After that, he claimed, Marjorie persuaded him against his better judgement to return to Canada. 'They don't like me in this country,' he confided to Jim Algie, the local *Sun Times* reporter who interviewed him in December 2003, adding with his usual tactless candour: 'I think the biggest mistake I ever made was coming back to this country in November '59. People said, "Oh, you live in Canada," and I said, "No, we exist here. We lived in Europe."'[1]

They went back to where his Waterton forebears had farmed near Owen Sound ('A deadbeat place. Very backward compared to the little city I grew up in [Camrose, Alberta]. But we liked the countryside.') Three years after his death Marjorie spoke of Bill's lasting love of England; but it must have been a complex affection since he left Algie in no doubt as to why he had given up all further hope of a flying career in the British aircraft industry. 'I got so pissed off with the people involved there. They were such bloody liars and crooks. And they were cutting corners, money, everything. As a test pilot you were sure as hell used. And they didn't pay much.' There was a strange moment in this interview when, given that Waterton seemed to regret having lived in Canada for the past half-century, Algie quite sensibly asked him: 'So why did you stay?' The answer was the single bleak phrase: '*In a rut.*'

A second son, John, was born in 1960 even though his father later confessed, 'I'm not much of a family man.' Waterton took a job as a warehouseman and on the side taught the local politician and businessman Eddie

Sargent to fly. Sargent was Mayor of Owen Sound and the warehouse's owner. For three years Bill flew him around to meetings in a succession of Cessnas. His log books tell the erratic tale. With the exception of two aerobatic flights in a Bücker Jungmeister biplane in 1958 (which must have taken him back pleasurably to his Gladiator days), Gloster's erstwhile chief test pilot didn't fly at all between 1956 and 1961. In that year he touchingly headed the page of his log book 'The Resurection' [*sic*] and for the next three years kept quite busy flying light aircraft as a pilot for hire. But after 1964 the man who had tamed the Javelin never flew again. 'I didn't *let* myself miss it,' he said. 'I would dearly have liked to have an aeroplane to fiddle around with but it just wasn't [financially] possible. It's no use hankering after something, is it? Once your cock stops working there's no point in worrying about women, is there?'

So for most purposes *in a rut* summed up the last half-century of his life. His tone of voice in the Algie interview is not bitter or recriminatory. It seems to look back on neither great pleasure nor great anguish. Rather, it reflects a kind of acceptance. Until 1956 he had done *that*, and since then he had been doing *this* – whatever it was, including growing older. In the face of such stoicism a biographer simply draws a blank for well over half his subject's lifetime, except to note that for a short period Waterton joined a welding class and left behind in Willy's shed some very strange sculptures he had made from black iron. They consist largely of jagged triangles set at all angles that might or might not represent swept-back or delta wings. It would probably be merely vulgar to read into them anything expressive of flight, let alone aspiration.

From outside, those fifty grounded years in rural Ontario might sound like a recipe for chronic boredom, coming as they did after two decades of varied, frantic, often life-threatening activity. Yet the idea that he might have been bored is firmly rejected by Richard Bentham, who was probably Bill Waterton's sole close friend for the last seven years of his life. Bentham is himself a retired test pilot who served for years in the Royal Canadian Air Force. While in the RCAF he flew the Avro Canada CF-100, the type that Waterton had test flown throughout its prototype's difficult and dangerous teething stages and which Bentham describes as the best operational all-weather fighter in North America in the late fifties. And he did mean all-weather. 'On the Canadian Pacific coast we would take off night after night into wild winter storms with nothing but angry ocean and mountains beneath. We hung on to the thin thread of radar coverage to get home. A few of our mates disappeared without trace off the face of the earth. Now that's lonely.'[2] In 1963 Bentham went to England to join the prestigious ETPS course, then at Farnborough, where at his very first lecture his tutor, Squadron Leader Eddie Rigg, declared: 'Gentlemen, you must never, *ever* forget that all aircraft manufacturers are thieves and rogues.' Waterton had not been mistaken.

One day in the late nineties Richard Bentham was at home some thirty miles south of Owen Sound when an elderly Oldsmobile rolled into his drive and an equally elderly driver emerged. It was Bill Waterton, who had learned from a mutual acquaintance that a former test pilot was living in the neighbourhood and had turned up out of the blue to talk. And talk he did, immediately and at length, Bentham recalls:

We hit it off and talked shop all afternoon. I knew of him and his CF-100 connection but not very much more. I had heard a few people piss and moan about his influence and mistakes. It was all nonsense. He was in his early eighties, vigorous, and seemed much younger. His memory for details – especially of the aircraft he'd flown – was razor-sharp. I could tell that he truly enjoyed swapping stories and so I made it a habit to call him about once a week or so for conversations that went on for an hour or more. He would call me if I were tardy. We got to know each other pretty well over the last few years of his life. He had a quick intelligence, a great sense of wicked humour, and placed a high value on good manners and civility.

I never thought of Bill Waterton as ever being bored. He had so many interests, such as his encyclopaedic knowledge of Second World War arms and ammunition as well as aircraft. He kept well abreast of current affairs and politics, in which his views were as staunchly defended as they were politically incorrect. And he always seemed to have projects and ideas on the go.

By the time Waterton died in 2006 Richard Bentham had become a considerable admirer, although not blinded to his friend's faults.

Yes, he could be ornery. Yes, he pissed off a large number of people. No, he probably wouldn't have been anybody's idea of an ideal father. Very determined and very stubborn on occasion. But a *very* interesting gent. Once, in a moment of exasperation at his refusal to allow himself to be nominated for the Canadian

Aviation Hall of Fame, I told him he was his own worst enemy. 'Oh yes,' he replied, 'I've been told that many times.'

There is no mistaking Bentham's admiration and fondness, and he was to write a lengthy appreciation of Waterton after his death for a British aviation magazine that in the event never published it. The belated *Globe and Mail* obituary quoted Bentham pointing out that Waterton was Canada's most internationally famous and accomplished test pilot and adding: 'No one else even approached his record of achievement. To this day, he remains virtually unrecognised in this country.'[3] When Waterton died there was at his request no funeral service, no notice and no religious observances. Richard Bentham gave a graveside oration in which he called his friend 'stalwart': sturdy and resolute, a man who would have been at home in a Victorian world of empire, honour, duty and courage. I suspect few who loved Bill would have quibbled with that, even though one can almost hear the gruff word 'Balls!' coming from beneath the freshly laid turf.

Maybe the praise that would have given him most satisfaction came from the British test pilot John Farley, who had been a classmate of Richard Bentham at ETPS in 1963 before going on to join Hawker Siddeley and spending nineteen years on the Harrier programme, to become that aircraft's acknowledged master demonstrator. In response to a posting on an internet forum in early 2003, Farley wrote that Bill Waterton had recently been made a member of the ETPS Association. He added: 'He was considered a bad trouble-maker back in the fifties because of his insistence on telling the truth about the aeroplanes he tested. Jeffrey Quill was his biggest fan – which says it all, really.* *The Quick and the Dead* was my bible when I was in the business and I still read the preface a couple of times a year – lest I forget what being a test pilot is actually about.'[4] When he heard of Waterton's death Farley wrote to Bentham, saying, 'I still use the preface to his book as an example to students of honesty and professional integrity (even if not combined with tact).'

If the exact half-century between Bill Waterton's disappearance from England and his death in Ontario in 2006 constitutes something of an enigmatic gap for a biographer, what is one to make of the steeply descending glide path described by British aviation over the same period? Looking away from the fallow skies of the twenty-first century and back to those crowded heavens of the fifties, the gloomy analyses of Bill Waterton and many of his informed contemporaries lose their immediacy and edge. Instead, pure nostalgia wells up to take their place: nostalgia for the many extraordinary aircraft the period produced. Of course, the latter-day invention of a post-war 'golden era' of British aviation is fraught with the charisma of longing, caught up as we are today in the paralysing romance of our own national failure. There is a double irony here. First, enough time has elapsed to halo the lens of hindsight. Even though it might not seem possible to some of us, the early Jet Age is in temporal

---

* Jeffrey Quill (1913–96) succeeded 'Mutt' Summers as Vickers's chief test pilot after test flying every single mark of the Spitfire, the aircraft with which he is famously associated.

terms already much further removed from the present day than it was from the Wright Brothers. This is deceptive, for Britain's many striking and original aircraft designs of the fifties (the English Electric Lightning, for example) still seem infinitely more modern to us than a Cody British Army Aeroplane No. 1 of 1907 would have done to someone at the end of the Second World War. The other irony is the implication that we are casting our gaze back to a time of much greater national strength and purpose. The reality, though, is that in many ways Britain in 1945 resembled an invalid who had miraculously survived mortal illness but who would never again regain her former strength.

Even so, for the next two decades the country maintained a vigorous, if grossly mismanaged, output of aircraft. If we now view those as 'golden years' (and many who were professionally involved at the time viewed them as no such thing) it is at least partly because Britain today lacks the political initiative, as well as the money and know-how, to support a thoroughgoing aviation industry. Once you have elected to get off a technological express train there is probably no way of ever catching it again. It has gone; and you are left behind, watching it recede while bravely consoling yourself with thoughts of your astuteness at having saved yourself the higher fare. The beautiful Swallows, Comets, Hunters, Vulcans, Lightnings (and yes, even the elegant Viscount) now stand as shining monuments to a former Britain. True, it was a Britain already in political decline; but the war's heroic afterglow still back-lit the indigenous aircraft that crossed our skies and impacted our fields in such gloriously uneconomic profusion.

One way of viewing the period might be as a grand swansong or coda to the process we Britons had ourselves started with the Industrial Revolution. The long, frequently brilliant chapter of mechanical inventiveness and manufacture that began with steam finally itself ran out of steam. This was not through any waning of either ingenuity or enthusiasm on the part of individuals, or even of the nation's aviation industry as a whole. It happened because, however unconsciously and blunderingly it was done, it became the policy of successive British governments to eradicate that industry as though it were an unruly wasps' nest by employing the slow cyanide of contradictory policies, the withholding of support and funds, and the progressive poisoning of morale. In fact, although not even the politicians themselves quite realised it – and certainly not at the time of the upbeat Festival of Britain in 1951 – this turned out to be merely part of a historic policy change to do away with *all* Britain's capacity as a serious industrial nation, abolishing not just a century of making its own cars but a thousand years of building its own ships. I suspect this policy was more unconscious than deliberately willed, and it is one whose consequences for the nation are still not fully apparent. It sounds improbable; yet there is surely no other interpretation to be made of the steady, decades-long demolition of the country's manufacturing capacity – including its most charismatic industry – other than that at some level it was absolutely intentional, no matter what lengths politicians went to in order to conceal this fact from both the electorate and themselves. What remains of Britain's aero industry today (essentially that near-unaccountable armaments multinational, BAE Systems – the 'monstrous octopus' *de nos jours* – plus Rolls-Royce and a handful of private defence contractors like QinetiQ) has nothing to do with Britain at the level of domestic patriotism, still less of defending the White Cliffs, and everything to do with trade. 'Of course!' one can hear today's young Elizabethans saying. 'What else?'

A realistic way to view this development would be to make the entirely reasonable point that aircraft have steadily become too complex and costly. No one but a superpower or the most dedicated nation can afford the ruinous expense of designing and building them on their own. For anybody else it is a matter of forming consortiums and sharing work and costs in joint projects. Even so, the sneaking thought remains that if it matters enough to a country in terms of its independence and the prestige of its national image, it can still keep the whole process in-house, albeit at enormous cost. (In Europe one thinks primarily of France and Sweden. Outside Europe diverse nations including Brazil, Indonesia, Japan, China, India and of course Israel maintain innovative and lively indigenous aircraft industries.) But ditheringly and in piecemeal fashion over the years, the UK decided it lacked the money and the will to support a self-sufficient aircraft industry to meet its defence needs and the export market. Yet if France still can, Britain probably could have, too. We chose not to, either because we had more urgent political priorities or because motives such as national prestige were deemed too weak or laughably outmoded. It might also be because enough people in Whitehall and Downing Street were in sympathy with the view expressed by Sir W. G. Nicholson, Chief of Imperial Staff (1908–12), earlier in the twentieth century when he delivered his notorious verdict: 'Aviation is a useless and expensive fad advocated by a few individuals whose ideas are unworthy of attention.'

Whatever the explanation, the fact remains that in the early twenty-first century the UK is no longer capable of producing an entire military aircraft or helicopter without help from abroad. Of that imposing and multi-talented – if chaotic – assortment of competing aircraft companies fifty years ago there remains virtually nothing today. One single BAE Systems factory at Warton puts together Eurofighters for an order that in any case may well dry up after 2014 and lead to its final closure. When it comes to manufacturing large aircraft, Britain's remaining expertise is in making the wings for Airbus airliners and for the A400M military transport. Here Britain can cheer itself with the thought that, according to the aviation journalist Richard Gardner, we are still 'the world leader in wing manufacturing'[5] – whatever that could possibly mean, given that foreign-built wings seem to hold the world's huge fleet of Boeings quite adequately aloft. But even such limited and dubious 'world leadership' has its price. 'If the UK aerospace manufacturing sector becomes totally dependent on work packages off programmes dictated by others then it will have no control over its destiny,' Gardner goes on to point out, adding with masterly understatement, 'Aerospace in the UK, unlike in most other major aerospace nations, does not enjoy a high political priority.'[6] If it did, one might guess that an effort would have been made to retain more than a faint echo of our former skill. Out of a mere handful of remaining small aircraft companies, Britten-Norman (part of B-N Group Ltd) is probably the largest, still producing and selling in limited numbers its outstanding Islander, Trislander and Defender series of rugged twin-engined aircraft. But they are no longer made in England. The airframes are built in Romania and shipped to Bembridge on the Isle of Wight for assembly. Of the great names in British aircraft components, most are either defunct or have been swallowed up by conglomerates, usually foreign-owned. Almost the sole exception is Martin-Baker, the world's longest-established manufacturer of ejector seats. The company is still managed by the sons of its founder out of its original headquarters at Denham, Bucks., and claims to supply ejector seats to over seventy per cent of Western air forces. A counter on the company's website is labelled 'Total Lives Saved', and as of writing this figure stands at 7,550.

Apart from such rare exceptions, and seemingly in belated fulfilment of Duncan Sandys's predictions, the sole indigenous aviation niche that shows much serious potential for growth is that of UAVs: the unmanned aerial vehicles increasingly used for reconnaissance and air strikes on ground targets. No doubt in terms of avionics they are massively high-tech, built as they are for remote operation, stealth and endurance rather than speed. Still, they utterly lack glamour. To an old New Elizabethan an aircraft without a pilot is not a proper aircraft at all, merely a radio-controlled model for grown-ups in uniform. It's no longer about *flying*, such as the twentieth century invented and revelled in. Such drones belong to an entirely grimmer, robotic age in which machines will increasingly be used to spy on, control and kill people ('interdiction' in Military Speak), even while flying thousands of miles away from their desktop controllers. Future air wars will involve neither gallantry nor heroism, just mouse-clicks at a safe distance.

The conventional question for a Briton to ask is: 'What went wrong?' The equally conventional response has been given in a thousand learned books and articles listing the UK aero industry's manifold faults and blunders. All they do is confirm in ever-greater scholarly detail the criticisms Bill Waterton made with such passion from ill-designed cockpits and in newspaper columns over half a century ago. A year after Waterton vanished from Britain Sir Roy Fedden published a book that delivered a scathing attack on how muddled government policy had now critically weakened the country's air power.[7] Fedden was one of the great names in aero engine development. He had co-founded the Bristol Aeroplane Co. in 1920 and was the firm's chief engineer for twenty-two years before serving as a special technical adviser to the Ministry of Aircraft Production during the war, for which he was knighted. He later took various government posts, eventually becoming a technical adviser to NATO. He was thus able to view the industry from a threefold perspective of engineering, politics and military strategy. His book with its unwelcome verdict was published (by a coincidence that could only have given it added force) in 1957 at the exact moment when it was announced, mere months after Duncan Sandys's White Paper, that the Fireflash, the guided missile

335

in which Britain had invested high hopes, would never now be used. It had cost the taxpayer £15 million.

Fedden was widely quoted in the press, and anyone who read a newspaper soon learned that since the war Britain had 'dabbled' in fifty-one different military and civil aircraft projects but without enough of the trained technicians who might have built them before they became obsolete. It was a saga of reckless squandering. Merely between 1952 and 1955, Fedden wrote, 'there was a criminal waste of technical manpower caused by eight [aero] firms being asked to tender for the same Operational Requirement, only to find, after eighteen months of concentrated work, that it was considered obsolete'. Still more recently, he said, the country had been working simultaneously on thirteen different fighter projects and eight different guided missiles, all of which was far beyond our capacity for R & D. We hadn't the technicians, the wind tunnels, the test equipment or computer resources. He cited the way we had embarked on four different V-bombers (counting the Short Sperrin) as a particular example of duplicated work that tied up our severely limited manpower, not to mention financial resources. As an engineer, Fedden saw Britain's future threatened by a chronic lack of engineers. 'We have the finest technicians in the world but ... there is not a single occupied chair of production engineering in any of this country's universities. Our fatal mistake has been to clutter up our industry and to attempt to compete on level terms with the other two great powers in military aircraft. We are suffering from delusions of grandeur.' Little change there in the last half-century, then. Still the nuclear pretensions and lack of engineers. No shortage of bankers, though.

So there it was: whatever else, a complete vindication of Bill Waterton's main charges against both the aviation industry and government policy that had been contained in his own book a year earlier. It is worth noting that after publishing his book Sir Roy was not hounded in his retirement or blackballed by his clubs. A distinguished Knight of the Realm with exalted political and military connections, he could air his opinion with impunity. A mere Canadian airman could not. Pilots, the Establishment evidently felt, should stick to flying aircraft and content themselves with aerobatics and glamour. As mere drivers they were distinctly 'other ranks' and ought not to have the temerity to take the officer class to task. It made no difference that Waterton had been a civilian since 1946. He had been testing military aircraft and was bound by the Official Secrets Act. He should have known his place and kept his mouth shut . . . One's admiration for the old boy's nerve goes up still further, and all the more so in comparison with his other test pilot colleagues who were to publish their memoirs – mostly much later and, with few exceptions, carefully steering clear of the political minefield.

As already mentioned, the 'Junior Service', as well as British aviation as a whole, has often exhibited a tendency towards conservatism, up to and including remarkably right-wing politics. Since this has been equally true of the country's executive echelons it is not easy to separate out how each played against the other when it came to reforming the aircraft industry. One should not underestimate the difficulty faced by the various British post-war governments, whether Conservative or Labour, when dealing with it. By the seventies, long after rationalisation and amalgamation had been forcibly imposed on the old aircraft companies and when it must have been obvious to even the densest aero executives that their entire sector's very existence was now under threat, old attitudes and behaviour patterns persisted to an extraordinary degree. Even ten years after Harold Wilson had conjured up his image of the white heat of technology, the mindset of a leisured business class could still effortlessly douse it in cold water.

Richard Bentham remembers a telling example of this. In late October 1970 he was in the UK test-flying a version of Short's twin turboprop Skyvan 300, an excellent light utility transport that is still flying in moderate numbers in the less accessible parts of the world. As before any test flight, he and the engineers had compiled a schedule of the different tests they needed to get through before coming down again. These things are planned methodically: time is of the essence. Flight-testing schedules usually have so many dependent financial ties and contracts attached that nothing – except possibly weather and safety issues – is allowed to stand in the way of getting the job done. On this particular morning Bentham was surprised when Short's man in the right-hand seat suddenly called a halt to the testing when they were little more than halfway through the schedule. 'Look at the time,' he said, tapping the Smith's clock on the instrument panel. 'If we're not careful we'll be late for lunch. That wouldn't do at all.'

In Bentham's experience it was unheard-of to interrupt a test schedule in order to eat lunch. Once a flight was over, grabbing a sandwich and a coffee was usually about as much as a test pilot could expect before he had to go up again or spend the afternoon writing his report. He landed the Skyvan with just enough time to change before being ushered into the directors' dining room for a long liquid lunch (unfortunately dry for working pilots): a full silver-service affair complete with several courses and wine waiters, port and brandy following afterwards. Around mid-afternoon the directors shuffled off, presumably for a nap. Somewhat incredulously, Bentham learned that this was no special occasion but simply how they habitually lunched at Short. As for the rest of the company, it ate in two separate dining rooms, the managerial 'officers' segregated from the blue-collar 'other ranks' who had their own canteen. That such a practice still existed at a time when British aviation was supposed finally to have become lean, mean and hungry for business came as a shock to Bentham. Still, he was Canadian and this was England, where they did things differently. To his amazement he later learned that a similar system of formal meals in three separate dining rooms still prevailed at that time even across the Atlantic at Avro Canada and de Havilland Canada. His surprise was identical to Stanley Hooker's over the 'Bristol lunches' that had so taken their toll of the working day back in the mid-fifties. The upshot for Bentham that afternoon in 1970 was that it took two flights and an extra day to do what could easily have been done in one. Despite this antique Top Table approach to management, Short, which was the world's oldest proper aircraft company (founded in 1908), survived until 1989 when it was acquired by the Canadian giant Bombardier Aerospace. It now makes aircraft components for its parent company and others at its Belfast factory.

Not that in those days a British aircraft company's fortunes could be held to ransom solely by its directors' dining traditions. At the other end of the social scale lurked the unions. In August 1977 Richard Bentham made three test flights from Hatfield with John Cunningham in the Hawker Siddeley (BAe/de Havilland) HS.125-700B ('a decent business jet, but the French Dassault Falcons were superior in every respect'). By this period in his career Cunningham had become something of a Grand Old Man among British test pilots, his wartime 'Cat's Eyes' image having been successfully trumped by his celebrated work for de Havilland on the Comet and latterly on the Trident. Only three years from retirement, Cunningham somewhat resembled a grog-blossomed choirboy, albeit an extremely affable and gentlemanly one. His speech and mannerisms struck Bentham the Canadian as 'typically English upper-class', none of which detracted from the man's skill and long experience as a chief test pilot.

'I recall John Cunningham's real concern about getting back on the ground half an hour before quitting time so that the day shift could get the airplane safely tucked into the hangar. Otherwise, as I remember it, for even a few minutes past the magic quitting time they would all get a full shift of overtime. Even the illustrious John Cunningham had no power here. I was amused – he was not. I never saw such a thing anywhere else in the world.'[8]

*

One ought always to be suspicious of those virtues on which nations most pride themselves. I grew up to inherit the ubiquitous popular wisdom that the one thing we British were good at was management and organisation which, together with our famous imperturbability, was how we Got Things Done even when we had our backs to the wall. This, of course, was by contrast with hysterical foreigners who – in a favourite unlovely metaphor – 'couldn't organise a piss-up in a brewery'. All too sober, we can now look back on the last sixty years as an awful demonstration of a leading industrial nation's absolute inability to organise an aviation industry: a failure we achieved with managerial skills that would have disgraced Noah, let alone a nineteenth-century mill owner. A notorious recent example of these same skills was afforded by the opening of Heathrow's Terminal 5 in early 2008. The short-lived scenes of chaos, with 500 flights cancelled and 23,000 passengers taking off without their luggage, were televised around the world.

That such criticisms of British managerial competence are not misplaced has been forcefully borne out recently by John Edgley, the chairman of AeroElvira. Edgley describes his entire career as one of continuous effort to further the cause of making light aircraft in the UK. He was the man behind the highly original Optica project: a radical design in the early eighties for a slow-flying observation light aircraft that was far ahead of its time when launched. The test pilot for the EA-7 Optica was Neville Duke (who, despite his back injury, went on test flying until 1994, when he was seventy-four). Today, a handful of EA-7s are flying in the UK, the US and Australia.

In a letter to *The Aerospace Professional* in 2010 Edgley gave as the principal reason for the Optica's failure to generate big sales: 'lack of manufacturing expertise, particularly in management. We simply seem to have no national system for training people in the necessary management skills.' He went on to admit that all the interest in setting up production of an updated version of the Optica was now coming from overseas. 'The interest from UK manufacturing is zero . . .'[9] Such official indifference surely lies at the heart of the country's waning economic power.

For a New Elizabethan who as a near-teenager saw the great Coronation fly-pasts of 1953 and can remember them as though it were yesterday, it is sometimes hard to get a grip on the slippery years between then and now. The gulf between that era's promise of national rebirth and the present-day reality is too great. Sometimes, in mulish mood, the New Elizabethan-turned-pensioner ponders whether, in terms of deciding where it wants to go and planning accordingly, his country may not have good claims to be consistently the worst-governed of any advanced industrial nation since the Second World War. There seems to have been so little shape or direction to any of it that one probably shouldn't single out the aviation industry for special lament. It all goes back to the question posed the day after VE day in 1945: *Now* what was Britain's role in the world to be? In the ensuing sixty years of Westminster debates and newspaper editorials this has never been satisfactorily answered. What ought we to be doing, militarily speaking? Should we be on our own? Are we just members of NATO? Are we part of Europe? Are we in league with the United States (even if the US patently couldn't care if we sank beneath the North Atlantic tomorrow)? Why do we need nuclear weapons? Just what on earth did we think we were doing in Afghanistan (of *all* places, given British history), fighting with our traditional gallantry on our traditional shoestring with shamefully inadequate back-up and equipment? Replies to all these are given in Parliament with equally traditional bluster and rhetoric, but the true answers remain unknown. Nothing further ahead than the next general election is ever envisaged. Still, the present dire lack of money means that the Strategic Defence and Security Review of October 2010 went further towards clarifying matters than any of its numerous predecessors did.

It is evident from the story of this post-war period that, nostalgia apart, it would be pure self-delusion to look back on it as a 'golden age' for British aviation, given that it marked the effective dissolution of the nation's aircraft industry. However, there were very definitely golden moments gleaming throughout, with golden men and golden machines that unquestionably blazed amid the dispiriting muddle. It was still an era of shirt-sleeved competence. The heroism of the pilots, the risk-taking bravery of the various companies and designers as well as the daring of the engineers: all deserve recognition. They stand in glorious contrast to the loss of nerve, vacillation and incompetence that characterised so many successive governments in their pretence of deciding whether Britain should continue to build aircraft even as they prolonged the charade of acting as though Britain were still a first-rate power. It scarcely matters now whether the decision to dismantle British aviation was actually taken or whether it was simply allowed to go by default. There is no going back. Just as there remains only a handful of Britons who still know how to make a beer barrel out of oak staves or lay a hedge properly, so there is virtually nobody left who knows how to design and build even a light aircraft from scratch. This may or may not matter in the long run – only time or some unimaginable national emergency will tell. But in the short run it feels like pure loss: the casual draining of a painfully acquired reservoir of national know-how that amounts to a form of treason.

If one were looking for a physical monument to that whole period when Britain possessed world-calibre industrial skills, one could do worse than consider the old National Gas Turbine Establishment at Pyestock, Farnborough. It was a 108-acre site of gigantic derelict industrial architecture: silent, dignified and redolent of past achievement. It was the outcome of the wartime Gas Turbine Collaboration Committee that was set up to co-ordinate information from all the companies and engineers working on Frank Whittle's W.2 jet engine as well as on other gas turbine projects. Whittle himself worked there; Concorde's engines were tested there; and in between, every jet aircraft the British aircraft industry ever built was a direct beneficiary of the research, knowledge and skill amassed at NGTE.

No longer wanted by the government, the site was privatised in 2002 as part of QinetiQ, after which it fell steadily into the decay of abandonment. As someone who worked there in its heyday reminisced sadly, 'Those were the days when you could be proud of something.'

Back in 1968 matters of pride were also central to a private gesture that had public impact when it hit the headlines. Although it was intended as a protest, it could also have stood as an unofficial and heartfelt farewell to British aviation and to the old spirit of the RAF it was bound up with. Ever since the Sandys White Paper morale in all the services had declined, but nowhere had it slumped lower than in the RAF. The first of April 1968 marked the fiftieth anniversary of the forming of the RAF from the old Royal Flying Corps. The RAF began making preparations for a fly-past over London to commemorate the day. Although in terms of age it was the Junior Service, ever since the Battle of Britain it had been the most dramatically visible of the three: the country's premier quick-reaction fighting force that even in peacetime continued to suffer a heavy toll of casualties.

As the day approached, however, it became clear that Harold Wilson's Labour government was planning no such fly-past and murmurs of disquiet were heard in RAF stations across the land. What made 1 April even more poignant was the announcement that on that day Fighter Command would cease to exist. It and Bomber Command would be subsumed into something called Strike Command – a dim title that in an era of mounting industrial unrest also had all the wrong overtones. No more Fighter Command? My grandmother would have had a fit.

At No. 1 (Fighter) Squadron, West Raynham, Flight Lieutenant Alan Pollock was senior operational flight commander. This was the RAF's (and the world's) oldest military air squadron, and Pollock felt passionately that the service's fiftieth birthday ought to be marked by something more memorable than a royal anniversary dinner and a few parades. Accordingly, he and his colleagues organised several celebratory leaflet raids on various RAF stations to be carried out using the Squadron's Hawker Hunters. Pollock had seen action in the Middle East and was no stranger to skip bombing runs, so this would be a combination of professional skill and old-fashioned RAF exuberance, no doubt including a traditional 'beat-up' of the airfields. On 4 April they were all invited to a party at No. 1 Squadron's pre-war home in West Sussex – Tangmere. The party itself was excellent and included notables such as Air Vice-Marshal 'Micky' Martin, who had flown his Lancaster (P for 'Popsie') to bomb the Möhne Dam in the Dambusters raid. Still, at the back of everybody's mind was the melancholy awareness that Tangmere was scheduled to close as an RAF station (which it did in October 1970). Of all the airfields made famous in the defence of London during the Battle of Britain, Tangmere was perhaps the most numinous, more so even than Biggin Hill, Kenley, West Malling, Hornchurch, North Weald or Northolt. 'Teddy' Donaldson had been stationed there in 1940 when in command of 151 Squadron. So had Douglas Bader in 1941, flying as a wing leader with 616 Squadron when he was shot down over France and taken prisoner for the duration. The immensely dangerous secret SOE (Special Operations

Executive) flights had been based there, ferrying agents to and from Occupied France by night in 'Teddy' Petter's black Lysanders. After the war, Tangmere was home to the Fighter Leaders School and the High Speed Flight in which Bill Waterton and Neville Duke had assisted Group Captain Donaldson in setting the world air-speed record. Later still, in the year of the Coronation, it was the base from which Duke had captured the new world record in his red Hawker Hunter. (Unknown to all in 1968, Donaldson would even be buried at Tangmere in 1992 when he died at the ripe age – for a pilot of that era – of eighty.) To many of the revellers on the night of the party – and especially those who had overdone the champagne – it was hallowed ground. To them, Tangmere *was* the RAF; and its impending closure had about it an air of finality that went beyond the mere redeployment of a squadron.

The next morning, Al Pollock and the others climbed into their Hunters to fly back to West Raynham. Overnight, Al had taken a decision he kept to himself. Since London lay directly on his course to Norfolk, he was determined to slip away from the others and carry out a solo defiant 'celebration flag-wave' of his own over the Houses of Parliament, Downing Street, the Ministry of Defence and – most particularly – the RAF Memorial on the Westminster Embankment. What his comrades didn't know about it couldn't harm their careers. Pollock was flying XF 442, a Hunter FGA.9, the ground-attack version of the type. The pan at Tangmere now became loud with the shrill hiss of Avpin starters and acrid with drifting gases as the five started their Avons. As their jet-pipe temperatures stabilised the men carried out the ritual pre-flight check: bang seat live and pins stowed, trim, fuel, flaps, instruments, oxygen, hood, harness and hydraulics. Then, having checked in on Tangmere's local traffic radio frequency, brakes off and taxi out to turn on to the runway: 'the last section of the last Hunters and fighters of the RAF to fly into and out of our nation's historically greatest fighter airfield', as Pollock later wrote in valediction.[10]

Immediately after take-off he gave the others the slip, quickly dropping to low level so his Hunter's camouflage would render it invisible from above as he sped across the Sussex countryside. On the way to London he paused briefly to beat up Dunsfold airfield – the home of Hawker, who had designed and built so many great aircraft, including the one he was flying. Two minutes later he was over the reservoirs just to the south of Heathrow. He joined the Thames and flew eastwards along it, banking low to follow its sinuous curves. 'The weather was one of those rare, perfect, 8/8 Gordon's-crystal-gin-clear days when all the colours shout out brightly . . . I swept round over Wandsworth, Battersea and Chelsea bridges, keeping a special eye open for any helicopters.'

Within seconds he was over the Houses of Parliament and Whitehall. His approach so far had been discreet: slow and comparatively quiet. Now, though, in order to keep a good tight circle over Westminster, Pollock had to open the throttle and make a lot more noise. The roar of his Rolls-Royce jet at almost rooftop level 'was perhaps what was really necessary at this juncture to wake up our MPs and remind other august figures, sitting chairbound at their ministerial desks below, that we still had a fighting Air Force, one small unit of which was celebrating its anniversary, despite the dead hand of government policy and the sickening cut-backs of previous years'. Three times he circled low, the noise interrupting a debate in the House of Commons. Then he levelled out over the Thames and dipped his wings past the RAF Memorial on the Embankment. Satisfied that he had made his point, Pollock glanced at the fuel gauge and decided to carry on eastwards along the river to Essex and then turn north to West Raynham. But as he crossed London Bridge, travelling at about 330 knots now the need for stealth was over, Tower Bridge suddenly loomed ahead through his windscreen.

'Until this very instant I'd had absolutely no idea that, of course, Tower Bridge would be there. It was easy enough to fly over it, but the idea of flying *through* the spans suddenly struck me. I had just seconds to grapple with the seductive proposition which few ground attack pilots of any nationality could have resisted. Years of fast low-level strike flying made the decision simple.' What else could any RAF pilot officially rated as 'Exceptional' be expected to do, flying a Hawker Hunter illegally low over central London on a sparkling morning in April? He had already burnt his boats: his career was almost certainly in tatters. He was thirty-two years old and the father of four, with the responsibilities of a breadwinning family man – but what the hell. There are moments when you just have to go for broke. 'There was considerable road traffic I could now see, including a red double-decker bus slowly lumbering across the famous double-basculed bridge from north to south.' Calculating his clearances with split-second accuracy, Al Pollock took his camouflaged Hunter through the bridge above the traffic in a blast of motion and sound that beat back from the iron girders and startled the living daylights out of a good few people. By the time they realised what they had seen, Pollock's Hunter was a dwindling speck passing Rotherhithe in a shimmering blur of exhaust. He was not the first pilot to have flown through the bridge, but he was the first ever to do so in a jet aircraft.

On the way home Al Pollock beat up RAF stations Wattisham, Lakenheath and Marham. Finally, with less than 400 pounds of fuel left, he carried out 'a rather hurried, inadequate, inverted run over the squadron hangars at RAF West Raynham before breaking downwind, punching down the gear and landing, with the brake parachute bobbing about contentedly behind...' Within the hour all hell had broken loose. Pollock was formally placed under arrest on his station. Media reaction was split between two camps. 'Hunter Ace – Hero or Hooligan?' one newspaper asked, plainly unwilling to commit itself. There was a good deal of support from members of the public, like the lady who wrote to the *Daily Express* saying, 'Please don't condemn or punish the dare-devil pilot who swept across London. It did me – and a lot of other people – a world of good. I shall always remember the feeling of pride as I thought of that chap in control of so much power, and it revived memories of those wonderful fellows who during the war fought for our survival.' For the other side, the 'authorities' made predictably unsmiling statements, like the Metropolitan Police spokesman who said, 'We do not regard this as a joke. It could have had serious consequences. There were pedestrians and vehicles on the bridge.' How frightful. Only, of course, it didn't have serious consequences – except for Flight Lieutenant Pollock.

A short while later his squadron (which he should have been leading) was detached to North Africa for operational weapons training. Recovering from pneumonia and what he sensed were complications induced by the drug he had been prescribed, he felt increasingly isolated and abandoned. In the next six weeks his statutory right to see his Air Officer Commanding-in-Chief was twice denied before the Under-Secretary of State for Air formally announced to the press that Pollock would not be tried by court martial but instead would be invalided out of the RAF – and this before the medical board had even been convened. He was given no option in the matter. 'I was told that if I did not accept the invaliding out, with the inducement of my small pension, my services could be dispensed with under a certain Queen's Regulation without any formal disciplinary action or come-back.' Naturally there was a subtext to all this. By not putting Pollock before a court martial, the authorities were ensuring that his reasons for the stunt would get no public airing or act as a focus for the support he was already getting from the public at large as well as from all three services.

The incident came at a particularly bad moment for the Wilson government, which anyway faced rebellion in the press. But in addition there was the so-called 'Westminster Plot'. Within a month of Pollock's exploit and even as he was still in limbo under arrest, the hugely influential owner of the Mirror Group Newspapers, Cecil King, instigated a meeting with Lord Mountbatten and Sir Solly Zuckerman, Harold Wilson's chief scientific adviser, at which he suggested the time was now ripe to overthrow the government and replace it with a temporary administration headed by Lord Mountbatten. Zuckerman left the room calling the idea 'treason', and the plot fizzled. But given the atmosphere of paranoia at No. 10 Downing Street it is easy to see why the government was not keen to have Pollock take the stand at a court martial, explain exactly why the services were feeling so aggrieved, and cut a sympathetic – even heroic – figure.

In due course Al Pollock embarked on a new career and memories of the incident faded. But it is hard now not to see him as sharing a character trait with Bill Waterton. Both could be self-destructive when an issue of principle was involved. It is a quality that feels wholly admirable these days, when even disgraced public officials feel no compulsion to resign or even to apologise. One can certainly imagine Waterton applauding Pollock's defiant gesture on behalf of the service they both loved. Indeed, it is slightly surprising that he never did something similar himself to draw attention to how the practices of the aero industry were failing the country. The man who was invited to beat up the centre of Paris in a Meteor could so easily have done the same to London under the spur of truculence, buzzing Whitehall before sailing through Tower Bridge inverted. But eventually he had no need to, because for two years as a journalist he had a public forum that Alan Pollock was denied, and could make his points forcefully enough in print.

Well, it was all a long time ago. In the last forty years Britain has changed out of all recognition. The majority of the population has no memories of war, and the gallantries of yesteryear have inevitably lost their immediacy. To many, the Battle of Britain might as well be Waterloo: no less (but no more) real than its digital simulation as a desktop game. We are left with the irony that Gloster's name for Britain's first jet fighter, the Meteor, should also describe the trajectory of the country's post-war aero industry: a short-lived phenomenon that dazzled even as it was breaking up, leaving a trail seared across one's mental retina and a fallout of dust. Out of that dust sift memories of the heart-stoppingly sleek and dangerous machines that once roared over the Home Counties, whose approaching moan or exhilarating thunder would cause us New Elizabethans to drop everything and dash outside, craning skyward.

Yet recalling those aircraft now – the Meteors and Hunters and Vulcans and Lightnings – is more than a mere essay in nostalgia. Some of the qualities of the Britons who made and flew them rubbed off on their younger aspirants who now, decades later, feel compelled to honour the deal before it is too late, and remind another generation of what they magnificently achieved. No one today should ever sit cocooned in the safety of an Airbus or a Boeing without being aware of the hundreds of gravestones that paved the way to that safety. Beneath those slabs lie the remains of mostly young men: the smashed, the charred, the decapitated, as well as the brick-filled coffins of those who simply disappeared or were vaporised on impact. To us who once saw many of them fly they are still vivid. Until we ourselves are broken up for scrap or buried they will always be there, pummelling an inward sky with sonic booms and leaving it indelibly smudged with smoke and contrails.

Postscript

In the year of the RAF's centenary there is a journalistic tendency (it recurs in milder form each September) to hark back doggedly to a war now almost eighty years in the past. This not only underrates the RAF's remarkable competence in the tense decades that followed the Battle of Britain, but overlooks the often brilliant – even futuristic – science at RAE Farnborough and elsewhere that lay behind its capabilities. The full story of the extraordinary range and ingenuity of British aerospace research since 1939 has yet to be told, not least because much of it remains classified, particularly from 1960 onwards.

The episode of the wartime Tizard Mission with its present to the United States of a tin trunk containing the cream of Britain's research is already well known, although perhaps not widely enough. In that first year of the war, with so much high-pressure research and development being carried out simultaneously in so many fields, Britain found itself unable to manufacture some of the more complex devices its scientists were inventing. With great reluctance it was decided to farm out some of this work. In September 1940, at the height of the Battle of Britain when the outcome was still uncertain, the radar scientist Sir Henry Tizard led the British Technical and Scientific Mission to Washington DC bearing extraordinary gifts to the determinedly neutral Americans. These consisted of several top-secret inventions. The idea was partly to take advantage of America's superior developmental and manufacturing capacities but also in the hope of swaying Congress out of its short-sighted neutrality. The latter ploy clearly failed since it took the Japanese attack on Pearl Harbor a full fifteen months later to induce the US to enter the war.

The information the Tizard Mission handed over was later described by an American historian as 'the most valuable cargo ever brought to our shores'. It included a vital breakthrough in nuclear research, the secrets of plastic explosives and much else besides. But the 'pearl beyond price' was the cavity magnetron that had been invented only months earlier by two scientists at Birmingham University. This was a copper disc little bigger than the pendulum of a grandfather clock that could generate microwaves of unheard-of power, enabling high-resolution radar as well as greater accuracy in anti-aircraft guns. Not only did it guarantee the superiority of Allied radar during the war, but its long-term commercial value proved incalculable since the cavity magnetron is still at the heart of every microwave oven sold around the world. In addition, the Tizard Mission turned over a complete set of blueprints of Frank Whittle's jet engine.*

---

* For a fascinating account of this extraordinary transaction, told from the American side, see Jennet Conant, *Tuxedo Park* (Simon & Schuster, 2002). It was largely the enthusiasm and private enterprise of the millionaire scientist Alfred Lee Loomis that ensured the magnetron was first examined on the other side of the Atlantic by precisely those few physicists who could immediately understand and exploit it.

In fact, the American company General Electric already had engineering connections with Whittle and had been shown his first engine, the W.1, in Derby in 1939. By 1941 the US Special Committee on Jet Propulsion had distributed Whittle's blueprints and was already sponsoring three separate jet engine projects from Westinghouse, Allis-Chalmers and GE.

The nuclear research referred to was the so-called Frisch–Peierls Memorandum, named after two of the three Birmingham University scientists who had worked out the critical mass necessary for a nuclear chain reaction. These were priceless data; and when British and Canadian physicists subsequently collaborated in the Manhattan Project it was formally agreed that after the war all parties would regularly exchange information and pool their atomic secrets. Then on 1 August 1946 President Truman signed into law the Atomic Energy Act – also known as the McMahon Act after its sponsor – which immediately banned the US from sharing any nuclear information with any other country. Britain's resentment of this betrayal only deepened with the reflection that in order to preserve the nation's 'top table' status it would now need to develop its own nuclear bomb alone, not to mention a suitable bomber to deliver it, adding yet more expense to perceived injury. Hence the hurried genesis of the V-bombers.

The case of the M.52 (see p. 84) also represented considerable generosity with British expertise. The M.52 project was accorded the highest security classification of Top Secret (virtually no one other than Churchill, Beaverbrook, some Ministry of Aircraft Production officials and the Miles factory team knew anything about it). Nevertheless, in the autumn of 1944 the Ministry arranged a trip to the heavily guarded Miles Aircraft factory for a number of visiting foreigners. These were Americans representing the United States Army Air Force (USAAF), the National Advisory Committee for Aeronautics (NACA) and Bell Aircraft Corporation. On MAP's instructions Miles gave the visitors all the information on the M.52 they wanted and answered all their questions. It was also agreed that in three weeks' time a Miles design team would make a reciprocal visit to the USA and a similar full exchange of information on American supersonic progress would take place. Within days this projected visit was summarily cancelled by the Pentagon.

The reason why the Pentagon reneged on the agreement is still not clear. The M.52's nominated test pilot, the late Eric 'Winkle' Brown, hazards that it was because the US would have been embarrassed by how little they actually had to show their visitors. At that time they were still behind the British and the Germans in jet development, and having just seen a British project for a supposedly transonic aircraft taking shape they had probably gone away from the Miles factory considerably overwhelmed. However, they had also acquired a mass of vital information for free. Brown suggests that it was precisely because they were behind in jet engine development that they opted for rocket power for their Bell X-1 that went supersonic in 1947. Rocket power is very much simpler; and after the collapse of Nazi Germany the Americans made sure they got key technicians (Wernher von Braun), data, plans and equipment from the German V2 programme. Yet the X-1's aerodynamics were quite as critical as its engine, and Brown concludes that it was 'a mistake to hand over the complete M.52 data in August 1944'.

By the 1960s Britain's own space programme was also in full flow and many accomplished scientists were recruited by RAE Farnborough, especially for various rocket programmes, all of which were destined to be abandoned. Ground-breaking work was done on ion propulsion for spacecraft, while 1963 saw the invention at RAE of a patented process for harnessing the tensile strength of carbon fibre: a now ubiquitous material initially pioneered for jet engine fan blades as well as for lightweight components of air- and spacecraft in general.

The fate of Britain's proven satellite-launching capability in 1971 has already been noted. Yet that was only one of many projects British post-war governments abandoned after brilliant and promising work, mostly from cold feet, lack of money or both. Maybe most ahead of its time was the British Aircraft Corporation's completed blueprint for a British space shuttle in the 1960s – years before US designers were thinking along similar lines. BAC's team was led by a remarkable engineer, the late Tom Smith, who had joined Gloster in 1948 and then English Electric, where under Freddy Page he was one of a team of four working on the Canberra before he played a leading role in developing the Lightning and TSR.2. In the early 1960s the Air Ministry issued a contract for BAC at Warton to study the idea of hypersonic speed (i.e. Mach 5 and above).

'We started by looking at things that were Concorde-ish in nature,' Smith recalled, 'and went on from there to high speed aircraft which would travel at Mach 12. We gradually realized that we could go from air-breathers, which would stay in the earth's atmosphere, into space.'[1] Out of this grew MUSTARD: the Multi-Unit Space Transport and Recovery Device, the design work for which was completed in 1965. This envisaged three re-usable, crewed, delta-winged, lifting-bodied aircraft launched together as a single unit. Two of the aircraft would act as boosters for the third, at altitude donating their excess fuel to the third that was destined for orbit. After that the two boosters would separate and return to land independently, exactly as the Space Shuttle would from 1981, leaving the third to continue into space. This idea had the advantage over the later American system of not needing a huge rocket booster that would simply fall into the sea when exhausted. 'MUSTARD was regarded as a suitable project for joint development by European aerospace companies, with a cost estimated to be around "20 to 30 times cheaper" than that of the expendable rocket [satellite] launch systems of the time. Unfortunately, as with so many other British inventions, the government of the day decided not to proceed. About three years after MUSTARD was cancelled, the Americans became interested in a reusable aircraft.'[2]

An equally remarkable scientist was the late New Zealand-born aeronautical engineer Darrol Stinton who, unlike Tom Smith, also became a fully fledged test pilot, completing the ETPS course at Farnborough in 1960. In 1968 he was employed by the MoD's Operational Requirements Branch to design a Wide Speed Range Aircraft that would be able to hover as well as fly at Mach 0.8 and deliver a cargo of weapons or troops 'in the field from alongside the colonel's tent'. This was two years after the first Harrier GR.1s had been delivered to the RAF when it seemed highly desirable to develop a cargo-carrying VSTOL aircraft that was very much faster than a helicopter. Two of Stinton's concepts were to be powered by large rotors designed by the National Gas Turbine Establishment at Pyestock, Farnborough. His designs were eventually not backed and in time, together with related data from British university and industrial sources, the dossier was handed over to the US where it became the basis for the early development of the American V-22 tilt-rotor Osprey from 1981.

Britain still builds satellites, if not their launch vehicles. Surrey Satellite Technology Ltd has long worked with the UK Space Agency while also now being part of Airbus Defence and Space. Prominent among ongoing British space projects is Skylon, a single-stage-to-orbit spacecraft designed by Reaction Engines Ltd and powered by that company's revolutionary hypersonic SABRE (Synergistic Air-Breathing Rocket Engine). The spark of British engineering is thus still very much alight. However, it remains a spark and it is essential that the campaign to promote STEM subjects (Science, Technology, Engineering and Mathematics) in British schools is not allowed to lose impetus in the usual hallowed fashion but leads to a regeneration of our former national creative genius.

Notes

Chapter 1: Death at Farnborough

1 The Test Flying Memorial Project, www.testflyingmemorial.com.
2 *Flight*, 19 September 1952.
3 Wing Commander J. A. Robinson, *'I Am Saluting You, Sir'* (Old Forge, 2009).
4 James Gilbert, *The World's Worst Aircraft* (Hobbs/Michael Joseph, 1975).
5 Tony Blackman, *Vulcan Test Pilot* (Grub Street, 2007), p. 39.
6 *Time*, 15 September 1952.
7 Blackman, op. cit., p. 127.
8 Brian Rivas and Annie Bullen, *John Derry* (Haynes, 2008), pp. 177-8.
9 'WHH' (http://homepages.ihug.co.nz/~russells/nzrafaa/sbac.html).
10 Neville Duke, *Test Pilot* (Grub Street, 2006), pp. 167-8.
11 Ibid., pp. 10-11.
12 W. A. Waterton, *The Quick and the Dead* (Frederick Muller, 1956), pp. 155-6.
13 Rivas and Bullen, op. cit., pp. 123-4.
14 Ibid., p. 172.
15 Quoted in John Golley, *John 'Cat's Eyes' Cunningham* (Airlife, 2002), p. 203.

Chapter 2: Bill Waterton and the World Air-Speed Record

1 W. A. Waterton, interview with Jim Algie of the Owen Sound *Sun Times*, 11 December 2003.
2 W. A. Waterton, *The Quick and the Dead* (Frederick Muller, 1956), p. 25.
3 The Test Flying Memorial Project, www.testflyingmemorial.com.
4 John Collins, *Faith Under Fire* (Frewin, 1966), pp. 89-90, quoted in Martin Francis, *The Flyer* (Oxford University Press, 2008), pp. 190-1.
5 Hugh A. Halliday, *Not in the Face of the Enemy: Canadians Awarded the Air Force Cross and Air Force Medal, 1918-1966* (Robin Brass Studio, 2000).

6 Waterton, op. cit., pp. 33–4.
7 This and the preceding two quotations are undated and unattributed. They are all from Waterton's private scrapbooks but unfortunately cropped as to date and source.
8 Ditto, though probably the *Daily Telegraph*, circa 8 August 1946.
9 See Francis, op. cit., pp. 57–62.
10 Waterton, op. cit., p. 33.
11 Ibid., p. 41.

Chapter 3: The Sound Barrier

1 Quoted in John Terraine, *The Right of the Line* (Sceptre, 1988), p. 260.
2 Derek Wood, *Project Cancelled* (Bobbs-Merrill, 1975), p. 24, lists these as follows: **Airframe companies:** Airspeed Ltd, Sir W. G. Armstrong Whitworth Aircraft Ltd, A. V. Roe & Co. Ltd, Blackburn Aircraft Ltd, Boulton Paul Aircraft Ltd, The Bristol Aeroplane Co. Ltd, The de Havilland Aircraft Co. Ltd, The English Electric Co. Ltd, The Fairey Aviation Co. Ltd, Folland Aircraft Ltd, General Aircraft Ltd, Gloster Aircraft Co. Ltd, Handley Page Ltd, Hawker Aircraft Ltd, Miles Aircraft Ltd, Percival Aircraft Ltd, Saunders-Roe Ltd, Short Brothers (Rochester and Bedford) Ltd, Taylorcraft Aeroplanes (England) Ltd, Vickers-Armstrong Ltd, (Supermarine) Vickers-Armstrong Ltd (Aircraft Section), Westland Aircraft Ltd. **Aero-engines:** Alvis Ltd, Armstrong Siddeley Motors Ltd, Blackburn Aircraft Ltd (Cirrus Engine Section), Bristol Aeroplane Co. Ltd, The de Havilland Engine Co. Ltd, Metropolitan-Vickers Electrical Co. Ltd, D. Napier & Son Ltd, Power Jets (Research & Development) Ltd, Rolls-Royce Ltd.
3 Ibid., p. 40.
4 *The Aeroplane*, 26 October 1956, p. 618.
5 Don Middleton, *Test Pilots* (Guild Publishing, 1985), p. 127.
6 Wing Commander J. A. Robinson, *Jet Bomber Pilot* (Old Forge, 2006), p. 8.
7 Eric Brown, *Miles M.52* (History Press, 2012).
8 Tony Buttler, *Secret British Projects: Jet Fighters Since 1950* (Midland, 2000), p. 58.
9 Reginald Turnill and Arthur Reed, *Farnborough: The Story of RAE* (Robert Hale, 1980), p. 108.
10 See Andrew Nahum, 'The Royal Aircraft Establishment from 1945 to Concorde', in R. Bud and p. Gummett (eds), *Cold War, Hot Science: Applied Research in Britain's Defence Laboratories 1945–1990* (Science Museum, London, 2002), p. 56.

11 Wood, op. cit., p. 36.
12 Eric Brown, *Wings on My Sleeve* (Phoenix, 2007), p. 184.
13 *Flight*, 11 July 1946.
14 Brian Rivas and Annie Bullen, *John Derry* (Haynes, 2008), pp. 89–91.
15 Brown, op. cit., p. 157.
16 John Golley, *John 'Cat's-Eyes' Cunningham* (Airlife, 2002), p. 111.
17 Geoffrey Dorman, *British Test Pilots* (Forbes-Robertson, 1950), p. 22.
18 Roland Beamont, *The Years Flew Past* (Airlife, 2002), p. 102.

Chapter 4: A Risky Business

1 W. A. Waterton, interview with Jim Algie of the Owen Sound *Sun Times*, 11 December 2003.
2 W. A. Waterton, *The Quick and the Dead* (Frederick Muller, 1956), pp. 51–5 *seriatim*.
3 Ibid., p. 61.
4 *Tatler*, 24 September 1947.
5 Waterton, op. cit., p. 63.
6 Ibid., p. 72.
7 Wing Commander J. A. Robinson, *Jet Bomber Pilot* (Old Forge, 2006), pp. 5, 7.
8 National Archives, AVIA 5/35.
9 Jim Winchester (ed.), *The World's Worst Aircraft* (Amber Books, 2005).
10 Brian Rivas and Annie Bullen, *John Derry* (Haynes, 2008), p. 128.
11 Waterton, interview with Jim Algie.
12 Waterton, op. cit., p. 169.
13 Ibid., p. 185.

Chapter 5: Canberras, Hunters and Patriotism

1 Roland Beamont, *The Years Flew Past* (Airlife, 2002), pp. 64–5.
2 Quoted in Roland Beamont, *Fighter Test Pilot* (Patrick Stephens, 1987), p. 68.
3 Ralph Swift, 'Flying on 527 Squadron – Early 1950s', www.rafwatton.info/.
4 Wing Commander J. A. Robinson, *Jet Bomber Pilot* (Old Forge, 2006), p. 25.
5 D. Collier-Webb in Bernard Noble, *Properly to Test*, vol. 2 (Old Forge, 2004), p. 85.
6 Robinson, op. cit., pp. 25–8.
7 Peter Masefield and Bill Gunston, *Flight Path* (Airlife, 2002), p. 28.

8 Quoted in Beamont, *The Years Flew Past*, p. 85.
9 F. R. Banks, *I Kept No Diary* (Airlife, 1983), p. 198.
10 Eric Brown, *Wings on My Sleeve* (Phoenix, 2007), p. 215.
11 Neville Duke, *Test Pilot* (Grub Street, 2003), p. 180.
12 Ibid., p. 157.
13 Ibid., pp. 170, 171.
14 Ibid., p. 182.

Chapter 6: Crash Landings

1 W. A. Waterton, *The Quick and the Dead* (Frederick Muller, 1956), p. 197.
2 Ibid., p. 127.
3 Ibid., pp. 202–9, somewhat abbreviated and edited.
4 *The Citizen*, 30 July 1952.
5 Waterton, op. cit., p. 210.
6 Ibid., p. 216.
7 *Daily Mail*, 20 July 1956.
8 Tony Blackman, *Vulcan Test Pilot* (Grub Street, 2007), p. 51.
9 See Waterton in the *Daily Express*, 22 June 1954.
10 See Derek Collier-Webb, 'Tested and Failed: Gloster Javelin', *Aeroplane Monthly*, May–June 1998.
11 Wing Commander J. A. Robinson, *Jet Bomber Pilot* (Old Forge, 2006), p. 67.
12 Blackman, op. cit., p. 19.
13 Ibid., p. 190.
14 *Time*, 15 October 1956.
15 *Flight*, 25 October 1957.
16 *Flight*, 8 November 1957, pp. 755–6.
17 See www.thunder-and-lightnings.co.uk/vulcan/gallery3.html.
18 Andrew Brookes, *V-Force: The History of Britain's Airborne Deterrent* (Jane's, 1983), p. 35.

Chapter 7: 'Nothing like a hundred per cent aeroplane'

1 W. A. Waterton, interview with Jim Algie of the Owen Sound *Sun Times*, 11 December 2003.
2 *The Guardian*, 4 December 2004.
3 W. A. Waterton, *The Quick and the Dead* (Frederick Muller, 1956), p. 232.
4 Richard Bentham, personal correspondence, 24 October 2009.
5 *Daily Express*, 3 March 1955.

6 Ibid., 4 May 1955.
7 Ibid., 8 June 1955.
8 Ibid., 25 August 1955.
9 *Time*, 13 August 1956.
10 Derek Collier-Webb, 'Tested and Failed: Gloster Javelin', *Aeroplane Monthly*, May–June 1998.
11 Waterton, interview with Jim Algie.
12 *Daily Mail*, 30 October 1956.
13 *The Citizen*, 25 July 1956.
14 Peter Twiss, *Faster than the Sun* (Grub Street, 2005), pp. 14–18.
15 Ibid., p. 25.
16 Ibid., p. 146.
17 Derek Wood, *Project Cancelled* (Bobbs-Merrill, 1975), pp. 86–7.

Chapter 8: 'A power of no good'

1 Quoted in T. Hewat and W. A. Waterton, *The Comet Riddle* (Frederick Muller, 1955), p. 14.
2 Quoted in Till Geiger, *Britain and the Economic Problem of the Cold War* (Ashgate, 2004), p. 174.
3 Peter Masefield and Bill Gunston, *Flight Path* (Airlife, 2002), p. 213.
4 See House of Commons Debates, 20 December 1955, Hansard, vol. 547, col. 301W (Written Answer from Frederick Erroll).
5 Masefield and Gunston, op. cit., p. 238.
6 Ibid., p. 246.
7 Stanley Hooker, *Not Much of an Engineer* (Airlife, 2002), p. 128.
8 Masefield and Gunston, op. cit., p. 250.
9 Ibid., p. 257.
10 Robert Gardner, *From Bouncing Bombs to Concorde* (Sutton, 2006), p. 116.
11 Ibid.
12 Ibid., p. 136.
13 Ibid., p. 137.
14 Ibid., p. 138.
15 See James Hamilton-Paterson, *Seven Tenths* (Faber, 2007), p. 272.

Chapter 9: Fighter Jock Heaven

1 Roland Beamont, *The Years Flew Past* (Airlife, 2002), p. 158.
2 Quoted in Bernard Noble, *Properly to Test* (Old Forge, 2004), p. 131.
3 See www.lightning.org.uk/archive/0303.php.

4 Beamont, op. cit., pp. 161–2.
5 Don Middleton, *Test Pilots* (Guild Publishing, 1985), p. 239.
6 Stanley Hooker, *Not Much of an Engineer* (Airlife, 2002), p. 148.
7 Eric Brown, *Wings on My Sleeve* (Phoenix, 2007), p. 262.
8 Roland Beamont, *Fighter Test Pilot* (Patrick Stephens, 1987), p. 98.
9 Personal interview with John Farley, 9 January 2009.

Chapter 10: Not with a Bang

1 W. A. Waterton, interview with Jim Algie of the Owen Sound *Sun Times*, 11 December 2003, for this and the following quotations.
2 Richard Bentham, personal correspondence and interviews, 2009, for this and the following quotations.
3 Toronto *Globe and Mail*, 17 July 2006.
4 www.pprune.org/archive/index.php/t-80859.html.
5 *Aerospace International*, August 2009, p. 34.
6 Ibid., p. 32.
7 Roy Fedden, *Britain's Air Survival* (Cassell, 1957).
8 Bentham, personal correspondence, 22 June 2009.
9 *The Aerospace Professional*, March 2010, p. 4.
10 See *FlyPast* magazine, September 1981, for this and the following quotations.

Postscript

1 Quoted in Smith's obituary (*Daily Telegraph*, 1 November 2012) and referring to his article in *FLIGHT International*, 24 March 1966: 'Economic Space Transport: BAC's Proposal for a Low-cost "Orbital Transporter"'.
2 Ibid.

Bibliography

The following is a list of the principal books and articles consulted:

Anon., *London Airport* (HMSO, 1956)
Aris, Stephen, *Close to the Sun* (Aurum Press, 2002)
Banks, F. R., *I Kept No Diary* (Airlife, 1983)
Barker, Ralph, *The Schneider Trophy Races* (Chatto & Windus, 1971)
Beamont, Roland, *Fighter Test Pilot* (Patrick Stephens, 1987)
—— *The Years Flew Past* (Airlife, 2002)
Blackman, Tony, *Vulcan Test Pilot* (Grub Street, 2007)
Boot, Roy, *From Spitfire to Eurofighter* (Airlife, 1990)
Brookes, Andrew, *V-Force: The History of Britain's Airborne Deterrent* (Jane's, 1983)
—— *Vulcan Units of the Cold War*, Osprey Combat Aircraft no.72, 2009
Brown, Eric, *Wings on my Sleeve* (Phoenix, 2007)
Bud, R. and Gummett, p. (eds), *Cold War, Hot Science: Applied Research in Britain's Defence Laboratories 1945–1990* (Science Museum, London, 2002)
Buttler, Tony, *British Secret Projects/Jet Fighters Since 1950* (Midland, 2000)
—— *British Secret Projects/Jet Bombers Since 1949* (Midland, 2003)
Carter, Nick, *Meteor Eject* (Woodfield, 2000)
Collier-Webb, Derek, 'Tested and Failed: Gloster Javelin', *Aeroplane Monthly*, May–June 1998
de Havilland, Geoffrey, *Sky Fever* (Airlife, 1999)
Dorman, Geoffrey, *British Test Pilots* (Forbes-Robertson, 1950)
Duke, Neville, *Test Pilot* (Grub Street, 2006)
Eden, Paul and Moeng, Soph (eds), *The Encyclopedia of World Aircraft* (Silverdale Books, 2002)
Edgerton, David, *England and the Aeroplane* (Macmillan Academic, 1991)
Farley, John, *A View from the Hover* (Flyer Books, 2008)
Fedden, Roy, *Britain's Air Survival* (Cassell, 1957)
Francis, Martin, *The Flyer* (Oxford University Press, 2008)
Gardner, Charles, *British Aircraft Corporation* (Batsford, 1981)
Gardner, Robert, *From Bouncing Bombs to Concorde* (Sutton, 2006)
Geiger, Till, *Britain and the Economic Problem of the Cold War* (Ashgate, 2004)

Gere, Edwin, *The Unheralded* (Andrus, 2008)

Gibbs, Philip, *The New Elizabethans* (Hutchinson, 1953)

Gibson, Chris and Buttler, Tony, *British Secret Projects/Hypersonics, Ramjets & Missiles* (Midland, 2008)

Gilbert, James, *The World's Worst Aircraft* (Hobbs/Michael Joseph, 1975)

Golley, John, *John 'Cat's-Eyes' Cunningham* (Airlife, 2002)

Green, William and Pollinger, Gerald, *The Aircraft of the World* (Macdonald, 1953)

Hadley, Dunstan, *Only Seconds to Live* (Airlife, 1997)

Halliday, Hugh A., *Not in the Face of the Enemy: Canadians Awarded the Air Force Cross and Air Force Medal, 1918–1966* (Robin Brass Studio, 2000)

Hayward, Keith, *Government and British Civil Aerospace* (Manchester University Press, 1983)

Hennessy, Peter, *Having It So Good* (Penguin, 2007)

Hewat, Timothy and Waterton, W. A., *The Comet Riddle* (Frederick Muller, 1955)

Hooker, Stanley, *Not Much of an Engineer* (Airlife, 2002)

Jackson, Robert, *Infamous Aircraft* (Pen & Sword Aviation, 2005)

Kynaston, David, *Austerity Britain* (Bloomsbury, 2007)

Lithgow, Mike, *Mach One* (Allan Wingate, 1954)

Masefield, Peter and Gunston, Bill, *Flight Path* (Airlife, 2002)

Middleton, Don, *Test Pilots* (Guild Publishing, 1985)

Nahum, Andrew, *Frank Whittle* (Icon Books, 2004)

Noble, Bernard, *Properly to Test* (Old Forge, 2004)

Page, Lewis, *Lions, Donkeys and Dinosaurs* (Arrow Books, 2007)

Payne, Richard, *Stuck on the Drawing Board* (Tempus, 2006)

Reed, Arthur, *Britain's Aircraft Industry* (Dent, 1973)

Rivas, Brian and Bullen, Annie, *John Derry* (Haynes, 2008)

Robinson, Wing Commander J. A., *Avro One* (Old Forge, 2005)

—— *Jet Bomber Pilot* (Old Forge, 2006)

—— *Tester Zero One* (Old Forge, 2007)

—— *'I Am Saluting You, Sir!'* (Old Forge, 2009)

Rossiter, Sean, *The Chosen Ones* (Douglas & McIntyre, 2002)

Shute, Nevil, *No Highway* (Pan, 1966)

—— *Slide Rule* (House of Stratus, 2000)

Terraine, John, *The Right of the Line* (Sceptre, 1988)

Tiratsoo, Nick (ed.), *From Blitz to Blair* (Phoenix, 1998)

Turnill, Reginald and Reed, Arthur, *Farnborough: The Story of RAE* (Robert Hale, 1980)

Twiss, Peter, *Faster than the Sun* (Grub Street, 2005)

Uttley, Matthew, 'Operation "Surgeon" & Britain's Post-War Exploitation of Nazi German Aeronautics', *Intelligence and National Security*, vol. 17, no. 2 (June 2002), pp. 1–26

Veale, S. E., *Guide To Flying* (Temple Press, 1942)
Vigors, Tim, *Life's too Short to Cry* (Grub Street, 2008)
Waterton, W. A., *The Quick and the Dead* (Frederick Muller, 1956)
Weir, A. N. C., *Verses of a Fighter Pilot* (Faber, 1941)
White, Rowland, *Phoenix Squadron* (Bantam, 2009)
Whittle, Frank, *Jet* (Pan, 1957)
Winchester, Jim (ed.), *The World's Worst Aircraft* (Amber Books, 2005)
Winfield, Roland, *The Sky Belongs to Them* (PBS, 1976)
Wolfe, Tom, *The Right Stuff* (Bantam, 1984)
Wood, Derek, *Project Cancelled* (Bobbs-Merrill, 1975)
Wood, Derek and Dempster, Derek, *The Narrow Margin* (Arrow, 1967)

Index